Lecture Notes in Economics and Mathematical Systems 409

Lecture Notes in Economics and Mathematical Systems

Arno Sprecher

Resource-Constrained Project Scheduling

Exact Methods for the Multi-Mode Case

Springer-Verlag
Berlin Heidelberg New York
London Paris Tokyo
Hong Kong Barcelona
Budapest

Author

Dr. Arno Sprecher
Christian-Albrechts-Universität zu Kiel
Institut für Betriebswirtschaftslehre
Olshausenstraße 40
D-24118 Kiel, FRG

ISBN 3-540-57834-X Springer-Verlag Berlin Heidelberg New York
ISBN 0-387-57834-X Springer-Verlag New York Berlin Heidelberg

CIP data applied for

Printed in Germany

Typesetting: Camera ready by author
SPIN: 10090269 42/3140-543210 - Printed on acid-free paper

Acknowledgements

During the years it took to finish this work a lot of people have supported me. First of all, I would like to thank my supervisor Prof. Dr. Andreas Drexl who initialized my work on the topic. Without his support, his helpful suggestions and comments this thesis could not have been finished. Especially his thorough knowledge of the field and the literature involved enabled me an easy start of the work on the subject.

I am thankful to my colleages of the Institute of Business Administration of the Christian-Albrechts-University in Kiel. Especially Knut Haase, Carsten Jordan and Rainer Kolisch made he[illegible]ul suggestions and comments on the subject of research. Moreover, I wish t[illegible]a Kantowski, who took care of the literature, Marc Schumann, who draw[illegible]graphics, Sönke Hartmann and Justus Kurth, who implemented some of th[illegible]ms.

I am also thankful to Uwe [illegible] from the PC-laboratory, who offered the technical support during the time the c[illegible]tational studies were performed.

Furthermore, I am grateful to J[illegible] Gerdes from the Insitute of Oceanographic Research who provided me with the opportunity to run a series of computational experiments on the machines of the institute.

Kiel, December 1993

Arno Sprecher

Preface

Within a project human and non-human resources are pulled together in a temporaray organization in order to achieve a predefined goal (cf. [20], p. 187). That is, in contrast to manufacturing management, project management is directed to an end. One major function of project management is the scheduling of the project. Project scheduling is the time-based arrangement of the activities comprising the project subject to precedence-, time- and resource-constraints (cf. [4], p. 170).
In the 1950's the standard methods MPM (Metra Potential Method) and CPM (Critical Path Method) were developed. Given deterministic durations and precedence-constraints the minimum project length, time windows for the start times and critical paths can be calculated.
At the same time another group of researchers developed the Program Evaluation and Review Technique (PERT) (cf. [19], [73] and [90]). In contrast to MPM and CPM, random variables describe the activity durations. Based on the optimistic, most likely and pessimistic estimations of the activity durations an assumed Beta-distribution is derived in order to calculate the distribution of the project duration, the critical events, the distribution of earliest and latest occurence of an event, the distribution of the slack of the events and the probability of exceeding a date.
By the time the estimates of the distributions have been improved (cf. e.g. [52] and [56]). Nevertheless, there are some points of critique concerning the estimation of the resulting distributions and probabilities (cf. e.g. [48], [49] and [50]).
If we additionally assume a non-deterministic structure of the project network, then the Graphical Evaluation and Review Technique (GERT) (cf. [80], [81], [92], [93]

and [122]) is used to obtain the results as described above. GERT-networks are especially used in R&D-projects where it is a priori difficult to forecast how the project has to be performed. That is, the activities are not realized with certainty but with a given probability.

The stochastic scheduling problems will not be dealt with in more detail. We consider another extension of MPM (CPM), the resource-constrained project scheduling. In contrast to MPM (CPM), resource-constraints are explicitly taken into account. Each activity has a fixed deterministic duration and requires certain amounts of the resources involved each period it is in process. The per period availability of the resources is limited.

In the multi-mode case the activities comprising the project can be executed in one of several ways called modes. Each mode represents a way of combining different resources and/or levels of resource requirements. That is, a resource-resource and a time-resource tradeoff can be realized.

In this thesis we will examine exact methods for solving the multi-mode resource-constrained project scheduling problem. The outline is as follows:

Chapter 1 presents the model we are dealing with. After the introduction of the resources involved the problem is described. Using critical path analysis bounds on the completion times are derived and subsequently used in the mathematical programming formulation.

Chapter 2 is a discussion of special cases of the model outlined. The versatility of the model is demonstrated. We transfer the flow-shop-, the job-shop-, the open-shop- and the assembly line balancing problem into a resource-constrained project scheduling problem.

Chapter 3 addresses variants and extensions of the model under consideration. The temporal constraints are generalized and the level of usage of renewable resources is considered to vary from period to period in which the job is active. Furthermore, different measures of performance are presented in order to judge the quality of a solution.

Chapter 4 is devoted to the definitions of different types of schedules. Based on the verbal descriptions of different types of schedules out of the shop-scheduling environment the definitions are formalized and adapted to project scheduling. Moreover, the definitions are illustrated by examples.

Chapter 5 presents an algorithm for solving multi-mode resource-constrained project scheduling problems. The underlying concept is the precedence tree which is firstly described. Then the algorithm for the minimization of the projects makespan is presented and later on extended to any regular measure of performance. The algorithm is modified by the use of priority rules and accelerated by bounding rules. The chapter finishes with some examples illustrating the limitations of the procedure.

Chapter 6 deals with the generation of problem instances. An algorithm using several project summary measures in order to obtain the project network, the resource demand and the resource availability is presented.

Chapter 7 presents the results of the computational studies. By the use of the project generator a broad range of instances has been generated and solved by the exact procedure proprosed. Moreover, we show the results of the related truncated exact method.

Chapter 8 is devoted to an artificial intelligence approach. Beside model reformulations we give a short introduction into CHARME, a logic programming language. Moreover, the results of the different model formulations in CHARME and a comparison with the standard problem solver LINDO is given.

Chapter 9 discusses the applications of the models and methods presented.

Chapter 10 states the conclusions and comments on future research including remarks on relaxations of the multi-mode resource-constrained project scheduling problem.

Contents

resources.

Chapter 1

The Model

We consider a project which consists of a set of activities. Equivalently we refer to the activities with the term job or task, where the latter one is more closely related to the operation within the machine scheduling environment (cf. [5]).
In contrast to the standard methods of project scheduling, MPM (Metra Potential Method) and CPM (Critical Path Method), the models and methods we are dealing with take care of resource requirements induced by performing a job. Therefore a distinction of different types of resources required for the completion of the project is presented in Section 1.1. Section 1.2 is then devoted to a more detailed discussion of the problem and the underlying assumptions. The multi-mode extension and the different types of resources are illustrated by an example. In Section 1.3 time windows for the completion times of the activities are derived by critical path analysis. The mathematical programming formulation of the problem is outlined in Section 1.4.

1.1 Resource Categories

Following the categorization scheme proposed by Slowinski (cf. [105], [106]) and Weglarz (cf. [120], [121]) we distinguish three types of resources required for the execution of the project. We have renewable, nonrenewable and doubly constrained

Renewable resources are available on a period-by-period basis , that is, the available amount is renewed from period to period (hour, day, week, month). The per period level may be constant or varying from period to period. For example manpower, machines, fuelflow and space are renewable resources.

In contrast to the renewable resources, nonrenewable ones are available on a total project basis, that is, instead of a limited per period usage of renewable resources we have a limited overall consumption for the complete project. This category of resources can for example be represented by money, energy and raw material.

A resource which is limited on per period basis as well as on a total project basis is called doubly constrained. Again money is an example if the per period cashflow and the total expenditures are limited. Manpower can be a doubly constrained resource, too, if for example a skilled worker can only spend a limited number of periods on the project.

1.2 Problem Description

We consider a project which consists of J activities. Due to technological requirements there are precedence relations between some of the jobs. A job j, $j = 2, \ldots, J$, is not allowed to be started before all its predecessors h, $h \in \mathcal{P}_j$, are completed. We represent the structure of the project by a so called activity-on-node network (AON) where the nodes represent the jobs and the arcs the precedence relations. We have the following assumptions on the structure of the network:

- Activity 1 is the only start activity (source) and activity J is the only finish activity (sink).
- The network is acyclic.
- The nodes of the network are numerically labeled, that is, a node j has always a higher number then all its predecessors.

Each activity j, $j = 1, \ldots, J$, may be accomplished in one out of M_j modes. These modes are non-preemptable, i.e. a job j once started in mode m is not allowed to be interrupted up to its completion. Performing job j in mode m takes d_{jm} periods and is supported by a set R of renewable, a set N of nonrenewable and set D of doubly constrained resources. Given an upper bound $\overline{T}$ on the projects makespan we have an available amount of K^{ρ}_{rt} (K^{δ}_{rt}) units of renewable (doubly constrained) resource r, $r \in R$ ($r \in D$), in period t, $t = 1, \ldots, \overline{T}$. The overall capacity of the nonrenewable resource r, $r \in N$, and doubly constrained resource r, $r \in D$, is given by K^{ν}_r and K^{δ}_r, respectively. If job j is scheduled in mode m then k^{ρ}_{jmr} units of renewable resource r, $r \in R$, are used and k^{δ}_{jmr} units of doubly constrained resource r, $r \in D$, are consumed each period job j is in process. Additionally k^{ν}_{jmr} units of nonrenewable resource r, $r \in N$, are consumed. The parameters are summarized in Table 1.1[1]. The objective under consideration is (e.g.) the minimization of the makespan.

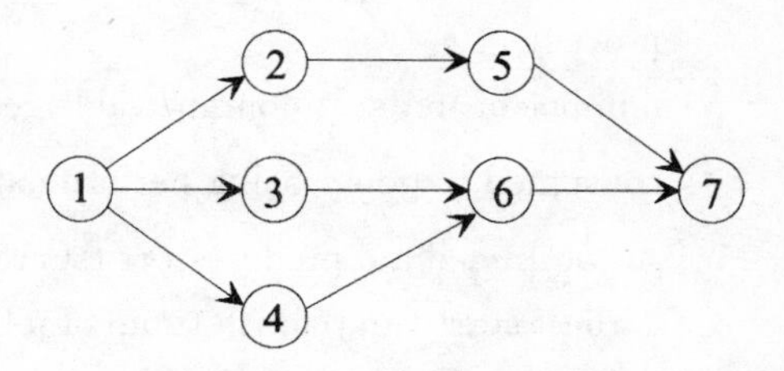

Figure 1.1: Example Network

We consider the example displayed in Table 1.2. The network of the project is given in Figure 1.1. Our project consists of seven activities ($J = 7$) and has an upper bound on the makespan of $\overline{T} = 48$ periods (days). All activities but activity one and seven can be performed in one out of three modes. The resources R_1 and R_2 represent two different machines, thus the per period (per day) availability is just one unit. Type R_3 reflects two workers with same capabilities, consequently two units of this resource are available per period. The nonrenewable resource N_1

[1]We assume all model variables and parameters to be integer-valued.

J	:	number of jobs
M_j	:	number of modes in which job j can be performed
d_{jm}	:	duration of job j being performed in mode m
R (N, D)	:	set of renewable (nonrenewable, doubly constrained) resources
$\overline{T}$	:	upper bound on the projects makespan
$K^{\rho}_{rt} \geq 0$ ($K^{\delta}_{rt} \geq 0$)	:	number of units of renewable (doubly constrained) resource r, $r \in R$ ($r \in D$), available in period t, $t = 1, \ldots, \overline{T}$
$K^{\nu}_{r} \geq 0$ ($K^{\delta}_{r} \geq 0$)	:	total number of units available of nonrenewable (doubly constrained) resource r, $r \in N$ ($r \in D$)
$k^{\rho}_{jmr} \geq 0$ ($k^{\delta}_{jmr} \geq 0$)	:	number of units of renewable (doubly constrained) resource r, $r \in R$ ($r \in D$), used (consumed) by job j being performed in mode m each period the job is in process
$k^{\nu}_{jmr} \geq 0$	:	number of units of nonrenewable resource r, $r \in N$, consumed by job j being performed in mode m
$\mathcal{P}_j$ ($\mathcal{S}_j$)	:	set of immediate predecessors (successors) of job j
ES_j (EF_j)	:	earliest start time (finish time) of job j, calculated by using minimal job durations and neglecting resource usage (consumption)
LS_j (LF_j)	:	latest start time (finish time) of job j, calculated by using minimal job durations, neglecting resource usage (consumption) and taking into account the upper bound $\overline{T}$ on the projects duration

Table 1.1: Symbols and Definitions

represents the money that can be spend on the project (26,500 units). For each job a mode dependent amount has to be paid for the execution. Speeding up the activities shortens the durations but increases the expenditures. The third mode is

J	$\overline{T}$			K^ρ_{1t}	K^ρ_{2t}	K^ρ_{3t}	K^ν_1	K^δ_{1t}
7	48			1	1	2	26,500	1
Requests and Durations								
Job	Successors	Mode	Duration	R_1	R_2	R_3	N_1	D_1
j	$\mathcal{S}_j$	m	d_{jm}	k^ρ_{jm1}	k^ρ_{jm2}	k^ρ_{jm3}	k^ν_{jm1}	k^δ_{jm1}
1	{2, 3, 4}	1	4	0	0	2	3,500	0
2	{5}	1	6	1	0	0	5,000	1
		2	6	1	0	1	5,000	0
		3	8	1	0	1	4,000	0
3	{6}	1	5	0	1	0	4,000	1
		2	5	0	1	1	4,000	0
		3	7	0	1	1	3,500	0
4	{6}	1	4	1	0	0	2,400	1
		2	4	1	0	1	2,400	0
		3	6	1	0	1	2,000	0
5	{7}	1	10	1	1	0	7,500	1
		2	10	1	1	1	7,500	0
		3	12	1	1	1	7,000	0
6	{7}	1	4	0	1	0	2,400	1
		2	4	0	1	1	2,400	0
		3	6	0	1	1	2,000	0
7	{}	1	5	0	0	2	1,500	0

Table 1.2: Example Problem

the "standard" way of execution which can be accelerated to the second mode, where additional money has to be spent. Furthermore, the doubly constrained resource D_1 represents an overload worker who is twenty periods available, that is, the overall capacity is $K^\delta_1 = 20$ and the per period availability is K^δ_{1t}. The overload worker can substitute any regular worker, i.e. one unit of the resource R_3. If we substitute a regular worker in the second mode by the overload worker we obtain the first mode.

1.3 Critical Path Analysis

In this section we derive time windows, i.e. intervals $[EF_j, LF_j]$, with earliest finish time EF_j and latest finish time LF_j, containing the possible completion times of activity j, $j = 1, \ldots, J$. The benefit is twofold: First, the number of variables used in the integer (binary) programming formulation is reduced substantially. Second, within a Branch and Bound algorithm the bounds can be efficiently used to speed up the convergence. If we have an upper bound $\overline{T}$ on the projects makespan, which can for example in the case of constant per period availability of the renewable resources be given by

$$\overline{T} := \sum_{j=1}^{J} \max_{m=1}^{M_j}\{d_{jm}\}, \tag{1.1}$$

we can calculate time windows for the completion times by making use of the precedence-constraints and modes of shortest duration. We assume the modes to be labeled with respect to non-decreasing duration and calculate by traditional forward and backward recursion the earliest finish times

$$EF_1 := d_{11} \tag{1.2}$$

$$EF_j := \max\{EF_i; i \in \mathcal{P}_j\} + d_{j1} \qquad j = 2, \ldots, J \tag{1.3}$$

and the latest finish times

$$LF_J := \overline{T} \tag{1.4}$$

$$LF_i := \min\{LF_j - d_{j1}; j \in \mathcal{S}_i\} \qquad i = J-1, \ldots, 1 \tag{1.5}$$

Consequently a job $j, j = 1, \ldots, J$, has to be finished within time window $[EF_j, LF_j]$. Note, since different modes may have different durations, starting an activity j within the analogously determined time window $[ES_j, LS_j]$ for the start times, means not necessarily that the job is completed in the interval $[EF_j, LF_j]$.

Figure 1.2 displays the results of the forward and backward recursion applied to the example presented in Section 1.2. Note, d_j reflects the shortest duration of an

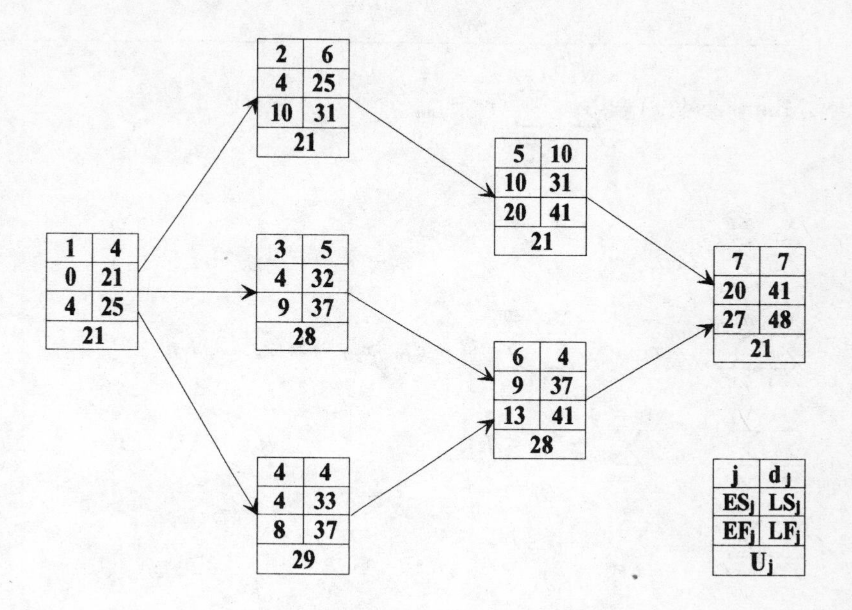

Figure 1.2: Results of the Critical Path Evaluation

activity j, $j = 1, \ldots, J$. U_j corresponds to the maximal delay job j can have without violating the upper bound on the makespan of the project.

1.4 Mathematical Programming Formulation

With the problem description given in Section 1.2 and the time windows derived in Section 1.3 we can now state the linear program, which was similarly presented by Pritsker et al. (cf. [94]), Bowman (cf. [15]) and Talbot (cf. [115]). We use binary decision variables x_{jmt}, $j = 1, \ldots, J$, $m = 1, \ldots, M_j$, $t = 1, \ldots, \overline{T}$,

$$x_{jmt} = \begin{cases} 1 & , \text{ if job } j \text{ is performed in mode } m \text{ and completed in period } t \\ 0 & , \text{ otherwise.} \end{cases}$$

The model is presented in Table 1.3. We refer to it as the generalized resource-constrained project scheduling problem (GRCPSP). The single-mode problem is

$$\text{Minimize } \Phi(x) = \sum_{m=1}^{M_J} \sum_{t=EF_J}^{LF_J} t \cdot x_{Jmt} \tag{1.6}$$

s.t.

$$\sum_{m=1}^{M_j} \sum_{t=EF_j}^{LF_j} x_{jmt} = 1 \qquad j = 1, \ldots, J \tag{1.7}$$

$$\sum_{m=1}^{M_h} \sum_{t=EF_h}^{LF_h} t \cdot x_{hmt} \leq \sum_{m=1}^{M_j} \sum_{t=EF_j}^{LF_j} (t - d_{jm}) x_{jmt} \qquad j = 2, \ldots, J, h \in \mathcal{P}_j \tag{1.8}$$

$$\sum_{j=1}^{J} \sum_{m=1}^{M_j} k_{jmr}^{\rho} \sum_{q=t}^{t+d_{jm}-1} x_{jmq} \leq K_{rt}^{\rho} \qquad r \in R, t = 1, \ldots, \overline{T} \tag{1.9}$$

$$\sum_{j=1}^{J} \sum_{m=1}^{M_j} k_{jmr}^{\nu} \sum_{t=EF_j}^{LF_j} x_{jmt} \leq K_r^{\nu} \qquad r \in N \tag{1.10}$$

$$x_{jmt} \in \{0, 1\} \qquad j = 1, \ldots, J, m = 1, \ldots, M_j, \; t = 1, \ldots, \overline{T} \tag{1.11}$$

Table 1.3: The Model of the GRCPSP

refered to as resource-constrained project scheduling problem (RCPSP). Note, for notational convenience, the RCPSP can be derived from the GRCPSP by skipping the mode index.

Since there is exactly one finish activity, the objective function (1.6) realizes the minimization of the projects makespan. Constraints (1.7) ensure that exactly one mode and one completion time is assigned to each activity . The precedence relations are taken into account by (1.8). For each renewable resource r, $r \in R$, on the left hand side of (1.9) the level of usage over the jobs which are active in period t, $t = 1, \ldots, \overline{T}$, is calculated. Thus (1.9) ensures that the per period availability levels are met. Taking into consideration that a job j, $j = 1, \ldots, J$, can only be completed within the time window $[EF_j, LF_j]$ we can rewrite (1.9) as follows:

$$\sum_{j=1}^{J}\sum_{m=1}^{M_j} k^{\rho}_{jmr} \sum_{q=\max\{t,EF_j\}}^{\min\{t+d_{jm}-1,LF_j\}} x_{jmq} \;\leq\; K^{\rho}_{rt} \qquad r \in R, t = 1,\ldots,\overline{T},$$

which substanially reduces the number of variables. (1.10) secures feasibility with respect to the consumable (nonrenewable) resources.

If doubly constrained resources have additionally to be taken into consideration, they can easily be incorporated in (1.9) and (1.10). If R, N and D are mutually disjoint sets we define

$$k^{\rho}_{jmd} := k^{\delta}_{jmd} \qquad d \in D \tag{1.12}$$

$$k^{\nu}_{jmd} := k^{\delta}_{jmd}\, d_{jm} \qquad d \in D \tag{1.13}$$

$R' := R \cup D$ and $N' := N \cup D$ and use R' and N' instead of R and N, respectively. The linear programming formulation suggests to employ a linear programming based algorithm, e.g. Branch and Bound with LP-relaxation, but the computational results in Section 8.3 will show that even small problems are intractable by this approach.

Chapter 2

Special Cases

In this chapter we will discuss some wellknown scheduling problems which turn out to be special cases of the more general resource-constrained project scheduling problem. More precisely, we will point out how to transfer the commonly known flow-shop- (FSP), job-shop- (JSP) and open-shop-problem (OSP) to a resource-constrained project scheduling problem RCPSP. Furthermore, by the reformulation of the assembly line balancing problem as a resource-constrained project problem the versatility of the model is illustrated.

The underlying idea for doing so is twofold: First, an algorithm which shows acceptable behaviour with respect to solution time and/or quality of the solution in dealing with the more general problem should be applicable with similar results in the special cases. Second, an idea for the theoretical or algorithmic treatment of the special problems, that is, in a more transparent context, might be generalized to a superior problem. Cf. e.g. the extension of the disjunctive graph concept which was originally developed for job-shop-problems (cf. [96]) and later on extended to single-mode resource-constrained project scheduling problems (cf. [7], [95]).

Further extensions and variants of the models under consideration as well as methods for dealing with the problems are thoroughly studied in [12] and [39].

2.1 Flow-Shop-Problem

We have a finite number A of orders, each of which has to be performed consecutively on the machines $R_1, \ldots, R_M$. Thus each order a, $a = 1, \ldots, A$, can be decomposed into M tasks $(a,1), \ldots, (a,m), \ldots, (a,M)$, where (a,m) indicates the processing of order a on machine R_m. Processing order a on machine R_m requires d_{am} periods, $d_{am} \geq 0$. Additional the following assumptions have to hold:

- A task (a,m) once started on machine R_m is not allowed to be interrupted.
- Each machine can process only one task at the same time.
- No order can be processed simultaneously on different machines.

Using the integer variables (cf. [40])

$$x_{amt} = \begin{cases} 1 & , \text{ if order a is completed on machine } R_m \text{ after } t \text{ periods} \\ 0 & , \text{ otherwise} \end{cases}$$

the problem can easily be stated as an integer program with the minimization of the lead time (makespan) as objective. Obviously we can calculate with

$$\overline{T} := \sum_{a=1}^{A} \sum_{m=1}^{M} d_{am} \tag{2.1}$$

an upper bound of the lead time, which enables us to calculate the time windows as presented in Section 1.3. We use the following redefinitions (cf. [40]):

- We introduce two dummy activities, activity 1 and activity $J = A \cdot M + 2$, which will be the source and the sink of the project network to develop.
- Each task (a,m), $a = 1, \ldots, A$, $m = 1, \ldots, M$, is assigned a job number j,

$$j := L(a,m) = (a-1) \cdot M + m + 1. \tag{2.2}$$

- The successors $\mathcal{S}_j$ of acivity j, $j = L(a,m)$, are calculated as follows:

$$\mathcal{S}_1 := \{L(a,1); a = 1,\ldots,A\} \tag{2.3}$$

$$\mathcal{S}_J := \emptyset \tag{2.4}$$

$$\mathcal{S}_j := \mathcal{S}_{L(a,M)} = \{J\} \qquad a = 1,\ldots,A \tag{2.5}$$

$$\mathcal{S}_j := \mathcal{S}_{L(a,m)} := \{L(a,m+1)\} \; a = 1,\ldots,A,\; m = 1,\ldots,M-1 \tag{2.6}$$

- We introduce a set R of renewable resources, $R := \{R_1,\ldots,R_M\}$, which correspond to the machines and have an availability of $K^{\rho}_{rt} = 1$, $r \in \{R_1,\ldots,R_M\}$, units each period t, $t = 1,\ldots,\overline{T}$.

- Since we have a single-mode problem, we can skip the mode index. We yield that processing job j, $j = L(a,m)$, $j = 2,\ldots,J-1$, takes $d_j := d_{am}$ periods, and uses

$$k^{\rho}_{jr} := \begin{cases} 1 & , \text{ if } r = R_m \\ 0 & , \text{ otherwise} \end{cases}$$

units. Furthermore, we define $d_1 := d_J := 0$ and $k^{\rho}_{1r} := k^{\rho}_{Jr} := 0$, $r \in R$.

2.2 Job-Shop-Problem

Again we have a finite number A of orders. Each order consists of G tasks, each of which has to be performed on one out of M machines $R_1,\ldots,R_M$, where different tasks use different machines. We assume $M = G$. In comparison to the flow-shop-problem, where every order has to be processed in the same ordering on the machines, the sequence is again prescribed, but may vary from order to order. Thus we have to illustrate a task by a triplet $(a, g, m(a,g))$, indicating that the g'th task of order a is processed on machine $R_{m(a,g)}$. Performing task g of order a on machine $R_{m(a,g)}$ requires $d_{a,g,m(a,g)}$ periods, $d_{a,g,m(a,g)} \geq 0$. The objective is again the minimization of the lead time and the assumptions given in Section 2.1 have to be taken into consideration.

For modelling purposes we introduce indices m_a, $a = 1, \ldots, A$, giving the number of the machine on which order a has to be finitely processed. Furthermore, integer variables x_{am}, $a = 1, \ldots, A$, $m = 1, \ldots, M$, are employed to represent the completion time of order a on machine R_m. The problem to solve is to sequence the orders on the machines, such that the maximum completion time of the orders is minimized. In order to meet the objective we use binary variables as proposed by Baker (cf. [5], pp. 206):

$$y_{abm} = \begin{cases} 1 & , \text{ if order } a \text{ precedes order } b \text{ on machine } R_m \\ 0 & , \text{ otherwise,} \end{cases}$$

$a = 1, \ldots, A$, $b = 1, \ldots, A$, $a \neq b$, $m = 1, \ldots, M$. Once the variables are defined the problem can easily be stated as an integer program.

Again we derive a single-mode resource-constrained project scheduling problem with serial network structure (cf. [40]):

- Two dummy activities, activity 1 and activity $J = A \cdot G + 2$, are introduced. They will be considered as the dummy start and dummy finish activity of the project network.

- Each task $(a, g, m(a, g))$, $a = 1, \ldots, A$, $g = 1, \ldots, G$, is assigned a job number j,

$$j := L(a, g, m(a, g)) := (a - 1) \cdot G + g + 1. \tag{2.7}$$

- The successors $\mathcal{S}_j$ of activity j, $j = L(a, g, m(a, g))$, are calculated as follows:

$$\mathcal{S}_1 := \{L(a, 1, m(a, 1)); a = 1, \ldots, A\} \tag{2.8}$$

$$\mathcal{S}_J := \emptyset \tag{2.9}$$

$$\mathcal{S}_j := \mathcal{S}_{L(a,G,m(a,G))} = \{J\} \qquad a = 1, \ldots, A \tag{2.10}$$

$$\mathcal{S}_j := \mathcal{S}_{L(a,g,m(a,g))} = \{L(a, g+1, m(a, g+1))\} \quad a = 1, \ldots, A, \; g = 1, \ldots, G-1 \tag{2.11}$$

- We introduce a set R of renewable resources, $R := \{R_1, \ldots, R_M\}$, which correspond to the machines and have an availabilty of $K^\rho_{rt} = 1$, $r \in \{R_1, \ldots, R_M\}$, units each period t, $t = 1, \ldots, \overline{T}$, with $\overline{T} := \sum_{a=1}^{A} \sum_{g=1}^{G} d_{a,g,m(a,g)}$.

- Since we have a single-mode problem, we can skip the mode index. We yield that the processing of job j, $j = L(a, g, m(a, g))$, $j = 2, \ldots, J-1$, takes $d_j := d_{a,g,m(a,g)}$ periods. Furthermore, we set $d_1 := d_J := 0$.

- We determine the number of units of resource r, $r \in R$, used by a job j, $j = L(a, g, m(a, g))$, $j = 2, \ldots, J-1$, as follows:

$$k^\rho_{jr} := \begin{cases} 1 & , \text{ if } r = R_m,\ m = m(a, g) \\ 0 & , \text{ otherwise} \end{cases}$$

and $k^\rho_{1r} := k_{Jr} := 0$, $r \in R$.

2.3 Open-Shop-Problem

We consider a finite number A of orders. M machines $R_1, \ldots, R_M$, are available to work off the orders. Each order has to be processed on all the machines. Processing order a on machine R_m takes d_{am} periods, $d_{am} \geq 0$. In contrast to the job-shop-problem, where the order of the machines to be visited is prescribed, we now have the freedom to choose any sequence. We can state the related single-mode resource-constrained project scheduling problem with a very simple network structure:

- Two dummy activities, activity 1 and activity $J = A \cdot M + 2$ are introduced, which again represent the dummy source and sink of the network.

- Each task (a, m), $a = 1, \ldots, A$, $m = 1, \ldots, M$, is assigned a job number j,

$$j \ := \ L(a, m) := (a-1) \cdot M + m + 1. \tag{2.12}$$

- The successors $\mathcal{S}_j$ of activity j, $j = L(a, m)$, are calculated as follows:

$$\mathcal{S}_1 := \{2, \ldots, J-1\} \tag{2.13}$$

$$\mathcal{S}_j := \{J\} \qquad j = 2, \ldots, J-1. \tag{2.14}$$

- We introduce a set R of $A + M$ renewable resources, $R := \{R_1, \ldots, R_{A+M}\}$, where $R_1, \ldots, R_A$ and $R_{A+1}, \ldots, R_{A+M}$, correspond to the orders and the machines, respectively. All of them have an availability of $K^\rho_{rt} = 1$, $r \in R$, units each period t, $t = 1, \ldots, \overline{T}$, with $\overline{T} := \sum_{a=1}^{A} \sum_{m=1}^{M} d_{am}$.

- Again we a have a single-mode problem, where processing of job $j = L(a, m)$ takes $d_j := d_{am}$ periods, $j = 2, \ldots, J-1$. Additionally we set $d_1 := d_J := 0$.

- We determine the number of units of resource r, $r \in R$, used by job j, $j = L(a, m)$, $j = 2, \ldots, J-1$, as follows:

$$k^\rho_{jr} = \begin{cases} 1 & , \text{ if } r = R_a \text{ or } r = R_{A+m} \\ 0 & , \text{ otherwise} \end{cases}$$

  and define $k^\rho_{1r} := k^\rho_{Jr} := 0$, $r \in R$.

Note, in contrast to the previously stated problems, where precedence relations prevent that an order a, $a = 1, \ldots, A$, is simultaneously processed on different machines, we now employ additional resources to enforce this. Every order a is assigned a resource R_a with an availability of one unit per period. Since every task of order a uses the resource R_a, simultaneously processing of order a on different machines is impossible.

2.4 Assembly Line Balancing

Whereas the problems considered in the previous sections are more closely related to small batch production with short or medium term horizon, the assembly line balancing problem addresses mass production with medium or long term horizon (cf. [47]). The problem is concerned with the assignment of tasks to (ordered) stations.

Given J tasks with associated processing times d_j, $j = 1, \ldots, J$, the work contents of a station g is calculated by the sum of processing times of the tasks assigned to station g. Thus the cycle time is given by the maximal work contents of the stations. Three objectives are most commonly considered (cf. [39], p.178, [75]): First, minimizing the production costs in the sense of the number of stations for a given output (production rate). Second, minimizing the cycle time, thus maximizing the production rate for a given number of stations. Third, minimizing the sum of the weighted cycle time and the weighted number of stations.

We consider the underlying assumptions of the first-mentioned problem more closely:

- Given the production rate, we calculate the cycle time τ.
- The production of one unit can be decomposed into J tasks, where task 1 and task J are the only start and finish tasks, respectively.
- Performing task j, $j = 1, \ldots, J$, takes d_j periods and is not preemptable.
- Due to technological requirements there are precedence relations between some of the tasks. They are imposed by sets $\mathcal{P}_j$, the set of tasks which immediately precede task j, $j = 1, \ldots, J$. The related graph is acyclic.
- Disjoint subsets of the set of tasks build the work contents of a station. The tasks out of each subset have to be performed consecutively within the cycle time.
- The objective is the minimization of the production costs, that is, the minimization of the number of stations.

Given an upper bound $\overline{G}$, $\overline{G} \leq J$, on the number of stations, we can introduce

$$\text{Minimize } Z(x) = \sum_{g=1}^{\overline{G}} g \cdot x_{Jg} \tag{2.15}$$

s.t.

$$\sum_{g=1}^{\overline{G}} x_{jg} = 1 \qquad j = 1, \ldots, J \tag{2.16}$$

$$\sum_{j=1}^{J} d_j \, x_{jg} \leq \tau \qquad g = 1, \ldots, \overline{G} \tag{2.17}$$

$$\sum_{g=1}^{\overline{G}} g \cdot x_{hg} \leq \sum_{k=1}^{\overline{G}} k \cdot x_{jk} \qquad j = 2, \ldots, J, h \in \mathcal{P}_j \tag{2.18}$$

$$x_{jg} \in \{0, 1\} \qquad j = 1, \ldots, J, \; g = 1, \ldots, \overline{G} \tag{2.19}$$

Table 2.1: Assembly Line Balancing Problem

binary variables x_{jg}, $j = 1, \ldots, J$, $g = 1, \ldots, \overline{G}$,

$$x_{jg} = \begin{cases} 1 & , \text{ if task } j \text{ is assigned to station } g \\ 0 & , \text{ otherwise.} \end{cases}$$

We obtain the mathematical programming formulation displayed in Table 2.1 (cf. [40], [85]). Since we have only one finish task the objective (2.15) realizes the minimization of the number of stations. (2.16) ensures that each task is assigned to exactly one station. By (2.17) we have, that the sum of the performance times associated with the tasks assigned to one station, does not exceed the cycle time τ. The precedence relations are taken into account by (2.18).

Note, we have assumed that the tasks assigned to one station have to be performed consecutively. Therefore, the precdence relations between a task h and a task j, $h \in \mathcal{P}_j$, that are not taken into account by assigning task h to a lower numbered station than task j have to be considered by first processing task h and then task j at the assigned station.

We can now reformulate the problem as a single-mode resource-constrained project scheduling problem, where the cycle time is reflected by a renewable resource:

- Each task j, $j = 1, \ldots, J$, is assigned an activity, the durations of which are given by the corresponding processing times.
- The precedence relations between the activites are the ones, induced by the one-to-one correspondence of the tasks and the jobs.
- We define $\overline{T} := \overline{G} \cdot (\tau + 1)$ and introduce a renewable resource R_1 with an availability of

$$K^{\rho}_{R_1 t} = \begin{cases} 0 & , \quad \text{if } t = g(\tau + 1), g = 1, \ldots, \overline{G} \\ 1 & , \quad \text{otherwise} \end{cases}$$

  in period t, $t = 1, \ldots, \overline{T}$.
- Each job j, $j = 1, \ldots, J$, uses one unit of resource R_1 each period it is in process.
- The objective is the minimization of the makespan of the newly derived project scheduling problem.

The essential clue is the time-varying supply. Since no unit of the renewable resource R_1 is available in the periods t, $t = g(\tau+1)$, $g = 1, \ldots, \overline{G}$, and, moreover, each job j, $j = 1, \ldots, J$, uses one unit of the resource R_1, it has to be performed in an interval $[(g-1) \cdot (\tau + 1) + 1, g \cdot (\tau + 1) - 1]$, $g = 1, \ldots, \overline{G}$. From an optimal solution of the related project scheduling problem one can easily derive an optimal solution of the assembly line balancing problem. The activities are assigned to the stations related to the intervals the activities are completed (performed) in.

Chapter 3

Variants and Extensions

In this section we deal with some generalizations of the resource-constrained project scheduling problem presented in Chapter 1. A discussion of a more general framework of precedence relations is given in Section 3.1. In Section 3.2 the time varying request for renewable resources is outlined. A brief discussion of further regular measures of performance follows in Section 3.3.

3.1 Generalized Temporal Constraints

Whereas in Chapter 1 we assumed that a job j, $1 \leq j \leq J$, is not allowed to be started before all its predecessors are finished, we will now generalize the temporal constraints. Given minimal and maximal time-lags between start and start, start and finish, finish and start or finish and finish of an activity h and an activity j (cf. Table 3.1) one can easily add the appropriate constraints.

We restrict ourselves to the minimal time-lag between the completion of an activity h and the start of an activity j, namely τ_{hj}^{min}, and the maximal time-lag between the start of an activity h and the completion of an activity j, i.e. $\hat{\tau}_{hj}^{max}$. Using the parameters and variables as presented in Section 1.2 and Section 1.4 we yield the

Time-lag between	Type	Symbol
start and start	minimal	$\tilde{\tau}_{hj}^{min}$
start and start	maximal	$\tilde{\tau}_{hj}^{max}$
finish and start	minimal	τ_{hj}^{min}
finish and start	maximal	τ_{hj}^{max}
finish and finish	minimal	$\check{\tau}_{hj}^{min}$
finish and finish	maximal	$\check{\tau}_{hj}^{max}$
start and finish	minimal	$\hat{\tau}_{hj}^{min}$
start and finish	maximal	$\hat{\tau}_{hj}^{max}$

Table 3.1: Time-Lags

new constraints:

$$\sum_{m=1}^{M_h} \sum_{t=EF_h}^{LF_h} t \cdot x_{hmt} + \tau_{hj}^{min} \leq \sum_{m=1}^{M_j} \sum_{t=EF_j}^{LF_j} (t - d_{jm})\, x_{jmt} \tag{3.1}$$

$$\sum_{m=1}^{M_j} \sum_{t=EF_j}^{LF_j} t \cdot x_{jmt} \leq \sum_{m=1}^{M_h} \sum_{t=EF_h}^{LF_h} (t - d_{hm})\, x_{hmt} + \hat{\tau}_{hj}^{max} \tag{3.2}$$

for $j = 2, \ldots, J$, $h \in \mathcal{P}_j$. The contraints (3.1) ensure that between the start of an acitivity j and the completion of an activity h, $h \in \mathcal{P}_j$, at least τ_{hj}^{min} periods have to pass, whereas (3.2) ensures that between the start of an activity h and the completion of an activity j, $h \in \mathcal{P}_j$, at most $\hat{\tau}_{hj}^{max}$ periods pass by.

Each precedence relation $h \in \mathcal{P}_j$ induces two constraints, one of the first and one of the second type. Clearly, one can skip (3.1) or (3.2) by defining $\tau_{hj}^{min} := -\infty$ or $\hat{\tau}_{hj}^{max} := \infty$, respectively. If the single-mode case is considered, then the different time-lags can be transformed to the standard type τ_{hj}^{min}. The transformations are summarized in Table 3.2 (cf. [7] and [38], pp. 88).

Note, in general the transformations given may lead to an acyclic network structure, which makes the resource-constrained project scheduling problems harder to solve as will be outlined in Section 5.6.

Given Time-lag	Transformation
$\tilde{\tau}_{hj}^{min}$	$\tau_{hj}^{min} = \tilde{\tau}_{hj}^{min} - d_h$
$\tilde{\tau}_{hj}^{max}$	$\tau_{jh}^{min} = -\tilde{\tau}_{hj}^{max} - d_j$
τ_{hj}^{min}	τ_{hj}^{min}
τ_{hj}^{max}	$\tau_{jh}^{min} = -\tau_{hj}^{max} - d_h - d_j$
$\check{\tau}_{hj}^{min}$	$\tau_{hj}^{min} = \check{\tau}_{hj}^{min} - d_j$
$\check{\tau}_{hj}^{max}$	$\tau_{jh}^{min} = -\check{\tau}_{hj}^{max} - d_h$
$\hat{\tau}_{hj}^{min}$	$\tau_{hj}^{min} = \hat{\tau}_{hj}^{min} - d_h - d_j$
$\hat{\tau}_{hj}^{max}$	$\tau_{jh}^{min} = -\hat{\tau}_{hj}^{max}$

Table 3.2: Transformations to the Standard Type

Inspite of this, by the introduction of more general temporal constraints some type of problems are reducible to easier ones (cf. [7]). E.g. the time-varying supply can easily be transformed to time constant supply, if dummy jobs are installed by time windows to use the non-existing overhead. In the single-mode case the time varying request (usage) problem as will be outlined in Section 3.2, can be represented by temporal constraints inducing some kind of parallelism of several jobs of different durations, all of which have a constant per period request.
Last but not least, job (project) specific release dates ρ_j and deadlines δ_j can be incorporated by defining the appropriate minimal and maximal time-lag between a dummy source activity and activity j.

3.2 Resource Requirements Varying with Time

The model outlined in Section 1.4 covers scheduling problems with time-varying availability, whereas the resource requirements are assumed to be constant. If we change to time-varying demand only the constraints on the limited per-period-usage have to be reformulated. We use the parameters introduced in Section 1.2 but instead k^{ρ}_{jmr} we use

$$k^{\rho}_{jmdr} \quad , \quad j = 1, \ldots, J, m = 1, \ldots, M_j, d = 1, \ldots, d_{jm}, r \in R,$$

where k^{ρ}_{jmdr} indicates the number of units of renewable resource r used by job j in mode m in the d'th period it is in process. The constraints (1.9) have to be changed to (cf. [43])

$$\sum_{j=1}^{J} \sum_{m=1}^{M_j} \sum_{q=t}^{t+d_{jm}-1} k^{\rho}_{jm,t+d_{jm}-q,r} \, x_{jmq} \leq K^{\rho}_{rt} \qquad r \in R, t = 1, \ldots, \overline{T}.$$

3.3 Further Regular Measures of Performance

In Section 1.4 we have used the makespan as performance measure, for later discussion it will pay to have a distinction on the performance measures under consideration. We extend the definition of regular performance measures given in [5], [54], [95] and [96] to the multi-mode case:

Definition 3.1
Let $\mathcal{M} := \{1, \ldots, M_1\} \times \cdots \times \{1, \ldots, M_J\}$ and $C_1, \ldots, C_J$ be the completion times of job 1,..., job J scheduled in mode $m_1, \ldots, m_J$, respectively. A performance measure Φ is a mapping:

$$\Phi : \mathbf{Z}^J_{\geq 0} \times \mathcal{M} \longrightarrow \mathbf{R}_{\geq 0}$$

which assigns to each pair of a J-tuple $C = (C_1, \ldots, C_J)$ of completion times and $M = (m_1, \ldots, m_J)$ of modes a performance value $\Phi(C, M)$. If Φ is monotonically increasing with respect to (the componentwise ordering of $\mathbf{Z}^J_{\geq 0}$ of) the first component, that is,

$$\Phi((C_1, \ldots, C_J), M) \; < \; \Phi((\overline{C}_1, \ldots, \overline{C}_J), M)$$

implies

$$C_j \; < \; \overline{C}_j$$

for at least one j, $j \in \{1,\ldots,J\}$, and additionaly minimization is considered, then we call the performance measure regular.

Clearly, we can denote Φ as a function of the binary variables introduced in Section 1.4, therefore, loosely in notation we yield the following regular measures of performance: Let ρ_j ($\overline{\delta}_j$) denote the release date (due date) of activity j. $j = 1,\ldots,J$.

(a) The minimization of the projects makespan

$$\begin{aligned}\Phi(C,M) &:= \sum_{m=1}^{M_J} \sum_{t=EF_J}^{LF_J} t \cdot x_{Jmt} \\ &= C_J.\end{aligned}$$

(b) The minimization of the weighted finish times

$$\Phi(C,M) := \frac{1}{J} \sum_{j=1}^{J} \sum_{m=1}^{M_j} \sum_{t=\overline{\delta}_j+1}^{LF_j} (t - \overline{\delta}_j) \cdot x_{jmt}$$

(c) The minimization of the total number of tardy activities

$$\Phi(C,M) := \sum_{j=1}^{J} \sum_{m=1}^{M_j} \sum_{t=\overline{\delta}_j+1}^{LF_j} x_{jmt}$$

(d) The minimization of the mean weighted flow time

$$\Phi(C,M) := \frac{1}{J} \sum_{j=1}^{J} \sum_{m=1}^{M_j} \sum_{t=EF_j}^{LF_j} (t \cdot x_{jmt} - \rho_j)$$

(e) The maximization of the net present value

$$\Phi(C,M) := \sum_{j=1}^{J} \sum_{m=1}^{M_j} \sum_{t=EF_j}^{LF_j} c_{jmt}\, x_{jmt}$$

where $c_{jmt} \geq c_{jm,t+1}$ for $j = 1,\ldots,J$, $m = 1,\ldots,M_j$, $t = 1,\ldots,\overline{T}-1$, denotes the cashflow induced by job j performed in mode m and completed in period t.

Note, more precisely a regular measure of performance is derived from the net-present value by multiplying it by -1.

Additional regular performance measures as e.g. the total (weighted) resource consumption and non-regular performance measures as the smoothness of the resource profile can be found in [107].

Chapter 4

Types of Schedules

In this chapter we state the formal definitions of different types of schedules. The original work can be found in [111], where beside the points to be explained here a set of examples brings the necessity of more formal definitions into focus. The schedules we are considering are the semi-active, the active and the non-delay schedules, where the latter one are only dealt with for the sake of completeness.

We restrict our considerations to the single-mode case, but the concept is simply adapted to the multi-mode case. The notation and symbols are derived from those of the GRCPSP of Chapter 1 by skipping the mode index. The problem obtained is the resource-constrained project scheduling problem (RCPSP). Clearly, nonrenewable resources do not have to be taken into account.

The outline of this chapter is as follows: In the first section we give a brief review of the literature dealing with the problem of schedule classification. The second section contains the formal definitions of different types of schedules. The belonging examples and illustrations are presented in Section 4.3. The sections correspond to the introduction and the equally named sections of the original work (cf. [111]).

4.1 Introduction

Classification of schedules is the basic work to be done in order to attack scheduling problems. For the case of the job-shop-problem (JSP) thorough studies have been performed (cf. [5], [22], [54] and [96]). Schedules for the JSP are classified as feasible, semi-active, active, and non-delay schedules. Procedures minimizing a regular measure of performance are usually enumerating semi-active or active schedules (cf. [96]). The latter are known to be the smallest dominant set of schedules (cf. [54], [96]). For the RCPSP as a generalization of the JSP the majority of researchers did not make use of any schedule classification (cf. e.g. [27], [88], [113], [115] and [116]).

Some researchers just defined the type of schedule needed. Thereby, different definitions have been proposed for the same type of schedules and identical definitions have been used for different kinds of schedules. E.g. Elmaghraby (cf. [49]) defines an eligible schedule as a schedule where no activity can be started earlier without changing the start times of any other activity and still maintain feasibility. In Schrage (cf. [101]) the same type of schedule is called active schedule. Wiest defines a left-justified schedule as a "feasible schedule in which ... no job can be started at an earlier date by local left shifting of that job alone" (cf. [123]) whereas Gonguet calls a schedule left-justified if "each job is scheduled as early as possible" (cf. [57]). Finally, other researchers have just taken over the schedule classification of the JSP without modifications (cf. [33], [34], [95]). This is somewhat reasoned by the way schedule classification is presented in the textbooks for the JSP most oftenly cited (cf. [5], [54]). There the definitions are more illustrative than formal and thus bear ambiguity in the case of the more general RCPSP.

In order to present a general and precise schedule classification we proceed as follows: Using the schedule classification for the JSP proposed by Baker (cf. [5]) as the stepping stone, we develop more formal and general definitions, that is, schedules for the RCPSP are discriminated to be feasible, semi-acitve, active, or non-delay schedules. Naturally, this classification holds for the JSP as well. Moreover, the (dominance) relations between the different sets of schedules, as known from the JSP, are preserved. This shall help to classify procedures for the RCPSP based on the schedules they examine.

4.2 Definitions

Recall, in the JSP (cf. Section 2.2) a number A of orders has to be processed on M machines. An order consists of G operations (tasks), each of which has to be performed on one of the M machines, $G = M$. It can be easily verified that the JSP corresponds to an RCPSP with $|R| = M$ renewable resources, each of which has an availability of one unit per period (cf. Section 2.2). As already mentioned we proceed as follows: Based on the informal definitions given in the JSP context (cf. [5]) we extend and formalize them for the RCPSP.

Within the JSP context Baker defines a schedule as a feasible resolution of resource and logical constraints (cf. [5], p. 179). More precisely we define as follows:

Definition 4.1

A schedule S is a J-tuple $S = (ST_1, ..., ST_J)$, where ST_j denotes the start time of activity j, $j = 1, \ldots, J$.

Definition 4.2

For a given schedule S and a period t, $1 \leq t \leq \overline{T}$, the set of activities being in progress in period t is $A_t(S)$,

$$A_t(S) \quad := \quad \{j; 1 \leq j \leq J, ST_j + 1 \leq t \leq ST_j + d_j\}.$$

Definition 4.3

A schedule S is called feasible, if the precedence relations are maintained, i.e.

$$ST_i + d_i \leq ST_j \qquad\qquad j = 2, ..., J, i \in \mathcal{P}_j$$

and the resource constraints are met, i.e.

$$\sum_{j \in A_t(S)} k_{jr}^{\rho} \leq K_{rt}^{\rho} \qquad r \in R, t = 1, \ldots, \overline{T}.$$

The local or limited left shift for a given schedule S of the JSP is defined as follows (cf. [5], p. 181): A local or limited left shift is, "moving an operation block to the left on the Gantt chart while preserving the operation sequences". Since the term "operation sequence" is not interpretable within the project scheduling context, we define:

Definition 4.4 (cf. [123])
A left shift of activity j, $1 \leq j \leq J$, is an operation on a feasible schedule S, which derives a feasible schedule S', such that $ST_j' < ST_j$ and $ST_i' = ST_i$ for i, $i = 1, \ldots, J$, $i \neq j$.

Remark 4.1
If a regular measure of performance Φ is considered and a schedule S' is obtainable from S by a left shift of an activity j, $1 \leq j \leq J$, then S is dominated by S' w.r.t. Φ.

Definition 4.5
A left shift of activity j, $1 \leq j \leq J$, is called a one-period left shift, if we have $ST_j - ST_j' = 1$.

Definition 4.6
A local left shift of activity j, $1 \leq j \leq J$, is a left shift of activity j which is obtainable by one or more successively applied one-period left shifts of activity j.

Remark 4.2
Within a local left shift each intermediately derived schedule has to be feasible by definition.

Regarding a schedule where no further local left-shifts are possible Baker defines a global left shift, as to start an operation earlier without delaying any other operation (cf. [5], p. 183). Instead we state:

Definition 4.7

A global left shift of activity j, $1 \leq j \leq J$, is a left shift of activity j, which is not obtainable by a local left shift.

Remark 4.3

(a) A global left shift of activity j, $1 \leq j \leq J$, induces $ST_j - ST'_j > 1$.

(b) If a feasible schedule S' is derived from the feasible schedule S by a global left shift, then S' is not obtainable from S by a local left shift, since at least one intermediate schedule is not feasible with respect to the resource constraints.

Based on the notion of a local left shift Baker defines the set of semi-active schedules, to be those schedules in which no local left shift is possible (cf. [5], p. 181). By employing our definition of a local left shift (Definition 4.6), we define:

Definition 4.8

A semi-active schedule is a feasible schedule, where none of the activities j, $1 \leq j \leq J$, can be locally left shifted.

Remark 4.4

A feasible schedule can be transformed into a semi-active schedule by a series of local left shifts. Note, in general, the derived semi-active schedule is not unique.

Obviously our definition coincides with the definition given by Baker for the JSP. The remark "In a semi-active schedule the start time of a particular operation is constrained by the processing of a different job on the same machine or by the processing of the directly preceding operation on a different machine" (cf. [5], p. 183) has to be generalized for the RCPSP in the following way:

Remark 4.5

In a semi-active schedule S the start time ST_j of any activity j, $1 \leq j \leq J$, cannot be reduced by one period, because there is at least one resource r, $r \in R$, for which the left over capacity in period $ST_j - 1$ is exceeded by the requirements of activity j or at least one predecessor of activity j is not finished up to the end of period $ST_j - 1$.

For the JSP the set of active schedules is defined as "the set of all schedules in which no global left shift can be made" (cf. [5], p. 183). For the RCPSP we use the following generalization:

Definition 4.9

An active schedule is a feasible schedule, where none of the activites j, $1 \leq j \leq J$, can be locally or globally left shifted.

Finally, in the JSP context (cf. [5], p. 185), a non-delay schedule is a schedule where "no machine is kept idle at a time when it could begin processing some operation". Employing the following remark we can give the more general definition of a non-delay schedule for the RCPSP.

Remark 4.6

Each RCPSP can be uniquely transformed into a unit-time-duration RCPSP (UTDRCPSP) where each activity j, $1 \leq j \leq J$[1], is split into d_j activities, each of which with duration one (cf. [27], [34]). Thus a feasible schedule S of the RCPSP uniquely corresponds to a feasible schedule UTDS of the UTDRCPSP.

Definition 4.10

A feasible schedule S for the RCPSP is called a non-delay schedule, if the corresponding schedule UTDS is active.

By definition we can state the following theorem:

Theorem 4.1

Let $\mathcal{S}$ denote the set of schedules, $\mathcal{FS}$ the set of feasible schedules, $\mathcal{SACS}$ the set of semi-active schedules, $\mathcal{ACS}$ the set of active schedules, and $\mathcal{NDS}$ the set of non-delay schedules, then the following holds:

$$\mathcal{NDS} \subset \mathcal{ACS} \subset \mathcal{SACS} \subset \mathcal{FS} \subset \mathcal{S}.$$

[1] If we have dummy acivities with a duration of zero periods then these activities obviously cannot be splitted.

4.3 Illustrations

In order to illustrate the given definitions we consider the example provided in Figure 4.1 with $|R| = 1$ and constant per period availability.

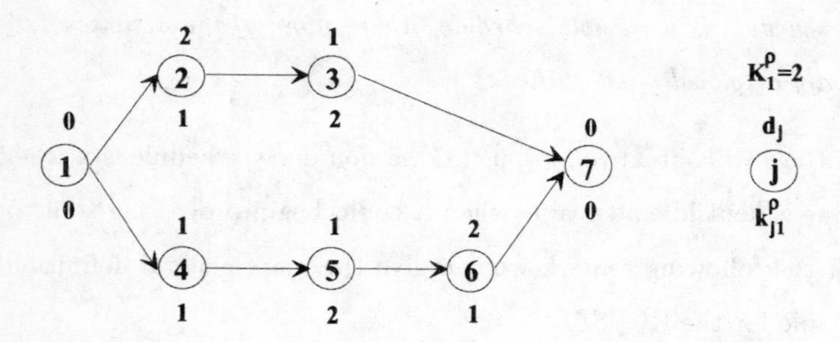

Figure 4.1: An Example for the RCPSP

A feasible schedule of the above problem is depicted by a Gantt-chart in Figure 4.2.

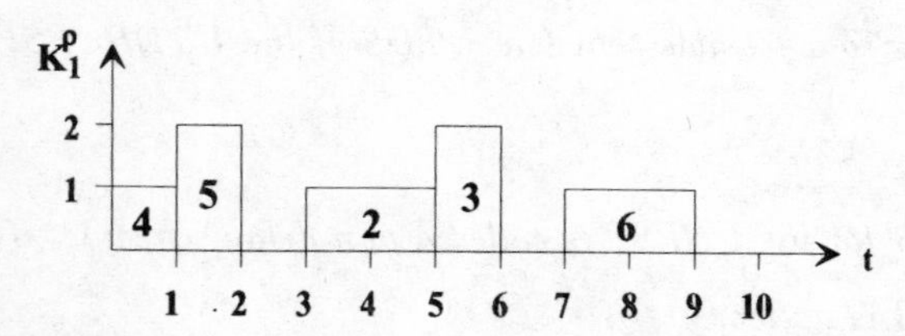

Figure 4.2: Feasible Schedule for the Example Problem

By performing a local left shift (constituting of a one-period left shift) of activities 2 and 3, respectively, and a local left shift (constituting of two one-period left shifts) of activity 6, the semi-active schedule displayed in Figure 4.3 is derived. Note that after the first one-period shift of activity 6 the intermediate schedule $S = (0, 2, 4, 0, 1, 6, 8)$ is feasible.

Regarding the semi-active schedule, clearly, none of the activities can be locally left shifted anymore. Nevertheless, activity 6 can be globally left shifted by performing

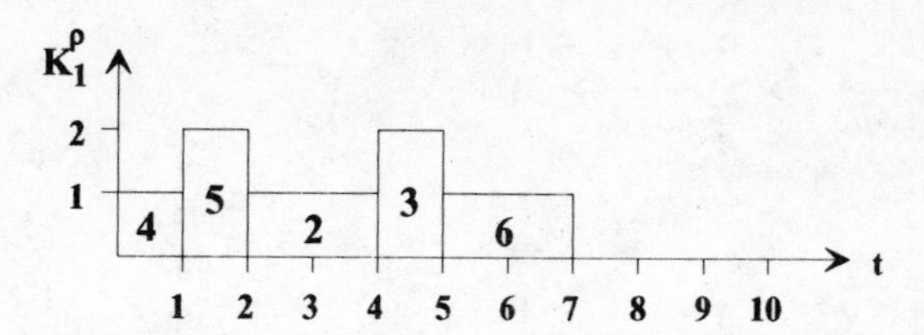

Figure 4.3: Semi-Active Schedule for the Example Problem

a three-period left shift. Doing so, one achieves the active schedule displayed in Figure 4.4.

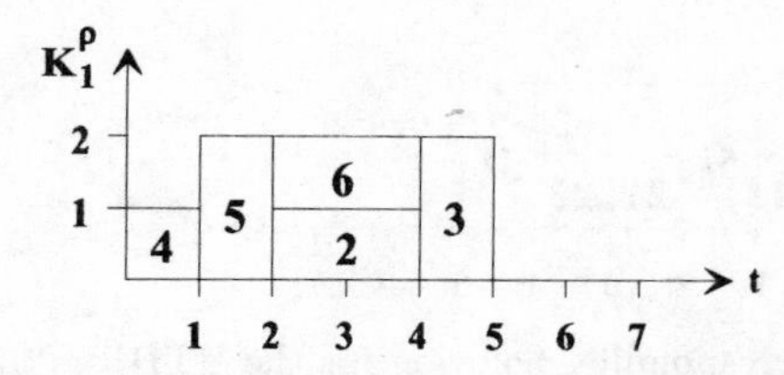

Figure 4.4: Active (and Unique Optimal) Schedule for the Example Problem

Note, the two intermediate schedules $S' = (0,2,4,0,1,4,6)$ and $S'' = (0,2,4,0,1,3,5)$ are not feasible, thus we have not performed a local left shift. Since none of the activities can be locally or globally left shifted, the schedule is active. Furthermore, the schedule is the unique optimal schedule of the example problem. Optimality can easily be verified by applying the resource-based lower bound as presented in [113], the uniqueness can be shown by performing an explicit enumeration with one of the schemes presented in [33], [113].

In order to see, whether the optimal solution is a non-delay schedule or not, we transform the example problem into the corresponding UTDRCPSP, where each activity j, $j = 1, \ldots, J$, of the RCPSP is transformed into the activites $j1, \ldots, jd_j$ (cf. Figure 4.5).

The solution corresponding to the one of Figure 4.4 is displayed in Figure 4.6. Since activity 21 can be globally left shifted ($ST_{21} = 2 \rightarrow ST'_{21} = 0$), the schedule for

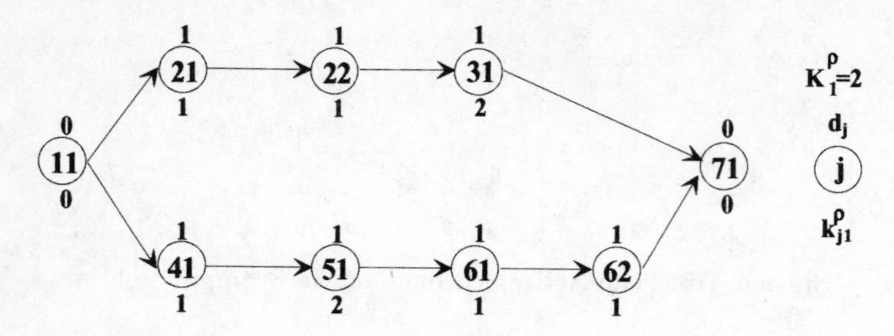

Figure 4.5: Corresponding UTDRCPSP

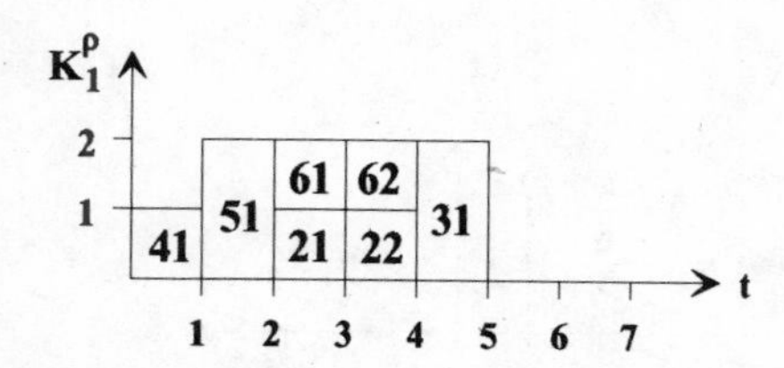

Figure 4.6: Corresponding Solution for the UTDRCPSP

the RCPSP does not belong to the set of non-delay schedules. Since the optimal schedule is unique, we can state the following: When considering a regular measure of performance, the set of non-delay schedules might not contain an optimal schedule.

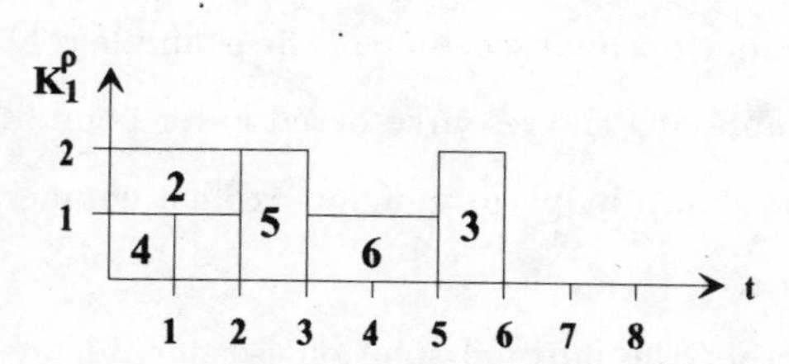

Figure 4.7: Non-Delay Schedule for the Example Problem

We obtain the schedule presented in Figure 4.7 from the one depicted in Figure 4.4 by assigning activity 2 the start time of 0, which alters the start times of the activites 3, 5 and 6. By Definition 4.10, the schedule derived is a non-delay schedule. Note, since the optimal solution is unique and not a non-delay schedule, we have an empty intersection between the set of optimal solutions and the set of non-delay schedules.

Chapter 5

A Branch and Bound Algorithm

Whereas exact methods for solving the single-mode resource-constrained project scheduling problem are well documented in the literature (cf. e.g. [7], [18], [25], [27], [32], [33], [33], [95], [101], [112], [113]), the multi-mode extension has attracted less attention (cf. [88], [89], [109], [114], [115], [116]). In this chapter we present and analyze an algorithm for the multi-mode resource-constrained project scheduling problem. The stepping stone of the algorithm is a proposal from Patterson et al. (cf. [88]) who introduced an enumeration procedure of the Branch and Bound (B&B) type. The basic ideas were, of course, given by Talbot and Patterson (cf. [116]) and Talbot (cf. [115]). For getting deeper insight the algorithm is completely restructured, where main differences will be identified by remarks.

We start the description of the algorithm in Section 5.1 with the precedence tree, which guides the search for a solution and is thus a dominant feature of the algorithm. Furthermore, an example illustrates that one of two interpretations of the relation between a father and his son within the precedence tree is wrong, that is, only heuristic solutions are obtained by the use of this interpretation. In Section 5.2 we describe the algorithm for the makespan criterion as objective function. The generalization of the Branch and Bound algorithm to any regular measure of performance is presented in Section 5.3. Section 5.4 is devoted to a brief discussion of priority rules. In Section 5.5 we present acceleration schemes which can easily be incorporated into the enumeration procedure and highly increase the performance of the algorithm. Section 5.6 points out the problems occuring with slightly changed assumptions. Especially the consideration of time-varying resource requests and maximal time lags will show that the algorithm has to be modified in order to work correctly.

5.1 The Precedence Tree

The main problem one encounters in the development of exact solution procedures of the Branch and Bound type is the construction of the enumeration tree. Roughly speaking the underlying idea is the successive decomposition of the problem into subproblems by splitting the feasible region and fixing variables. In some problems a part of the constraints can be easily incorporated into the contruction of the Branch and Bound tree, as it will turn out in the problem at hand.

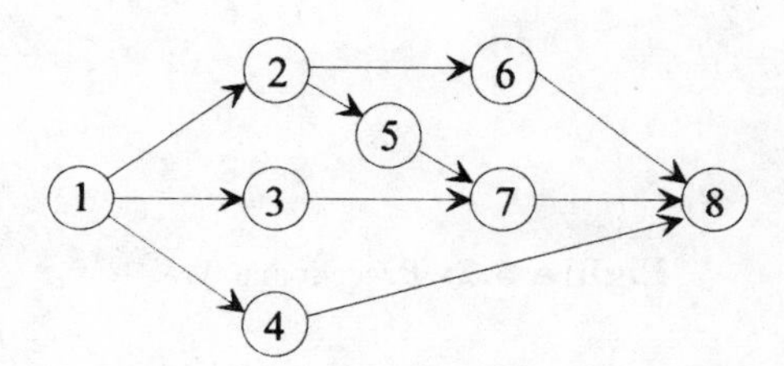

Figure 5.1: Example Network

For ease of notation we restrict ourselves to the single-mode case and consider the example given in Figure 5.1 (cf. [49], p. 179). Obviously activities 2, 3 and 4 cannot be started before activity 1 is finished. If activity 1 is scheduled activity 2, 3 and 4 become eligible. Thereby, an activity is called eligible if all its predecessors are scheduled.

The circumstances are illustrated in the related precedence tree (cf. Figure 5.2 (cf. [88], p. 12)). On every stage exactly one activity out of the set of eligible activities is scheduled. On the first stage only activity 1 is eligible and on the

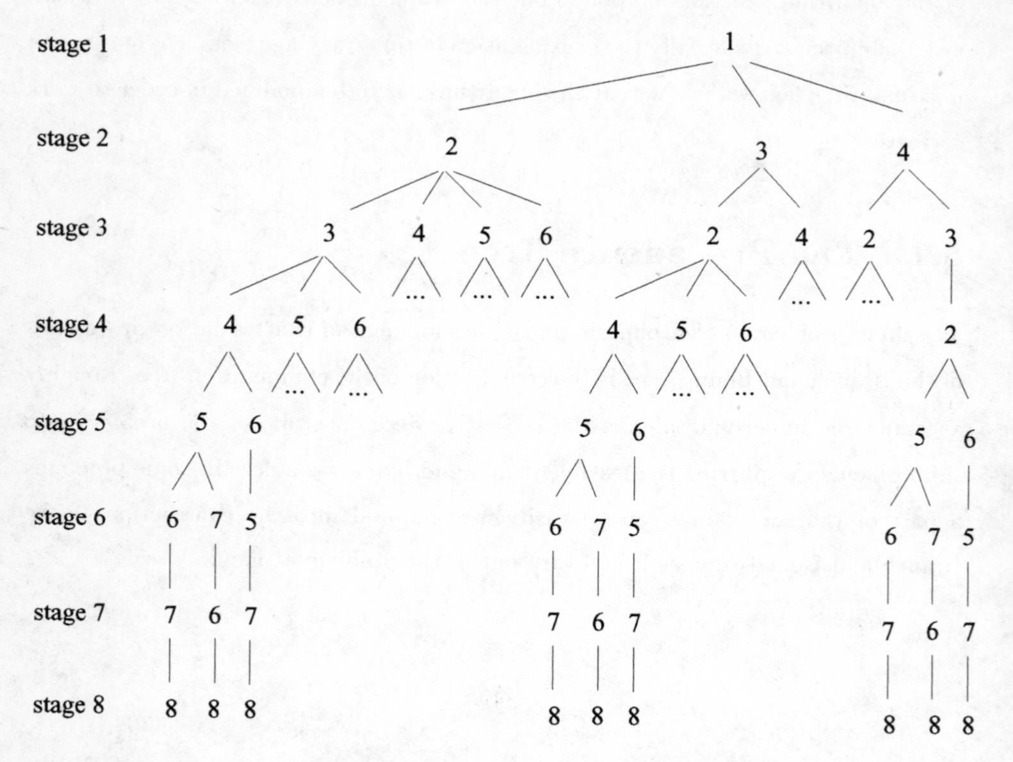

Figure 5.2: Precedence Tree

second (after activity 1 is scheduled) the activities 2, 3 and 4 become eligible. If we now schedule activity 2, then additionally activities 5 and 6 become eligible. If we schedule activity 3 or 4 on the second stage, then no additional activities become eligible. E.g. activity 5 and 6 are not eligible because not all their predecessors (activity 2) are scheduled.

Since we have exactly one start activity, we can successively determine the eligible set (set of eligible activities) Y_i of stage i, $i = 1, \ldots, J$. Of course, the sets Y_i are not independent of the "history", i.e. the set of already scheduled activities.

We denote with g_i the number of the activity scheduled on stage i and with $\mathcal{AS}_i$

the set of activities scheduled up to stage i. We can calculate[1]

$$Y_1 := \{1\} \tag{5.1}$$

$$\mathcal{AS}_1 := \{g_1\} = \{1\} \tag{5.2}$$

$$Y_{i+1} := Y_i \backslash \{g_i\} \cup \{k \in \mathcal{S}_{g_i}; \mathcal{P}_k \subset \mathcal{AS}_i\} \qquad i = 1, \ldots, J-1 \tag{5.3}$$

$$\mathcal{AS}_{i+1} := \mathcal{AS}_i \cup \{g_i\} \qquad i = 1, \ldots, J-1 \tag{5.4}$$

Remark 5.1

The essential limit (clue) is that not all the successors of the finally scheduled activity g_i become eligible on stage $i+1$. In the example above, if activity 3 is scheduled on stage 2 then none of its successors becomes eligible (cf.[88], p. 14).

The problem now arising is when to start activity g_i on stage i. Due to the precedence constraints activity g_i can only be started after the completion of all its predecessors, but, if there is no precedence relation between the activities g_i and g_{i+1}, there are two interpretations of the relation between father and son[2] within the precedence tree:

[1] Note, since we assumed that the network is numerically labled, activity 1 is always the unique eligible activity on stage 1.

[2] In the precedence tree (Figure 5.2) the activities 2, 3 and 4 are the sons of (the root) acitivity 1. Considering the branch 1-2, then activity 2 is the father of the activities 3, 4, 5 and 6.

Interpretation 1: The son (activity g_{i+1}) is not allowed to be completed before the father (activity g_i):

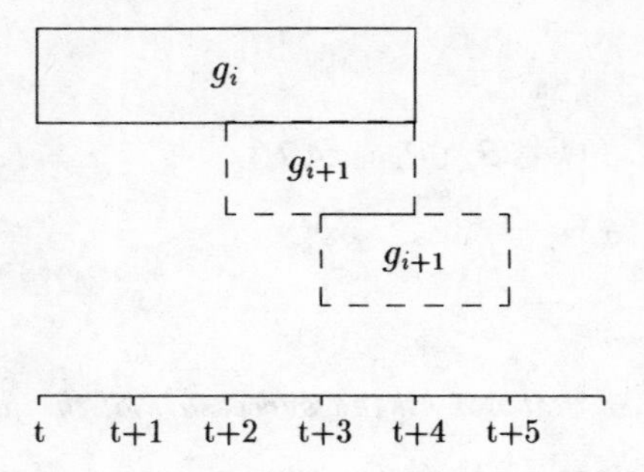

Figure 5.3: Illustration of Interpretation 1

Interpretation 2: The son (activity g_{i+1}) is not allowed to be started before the father (activity g_i):

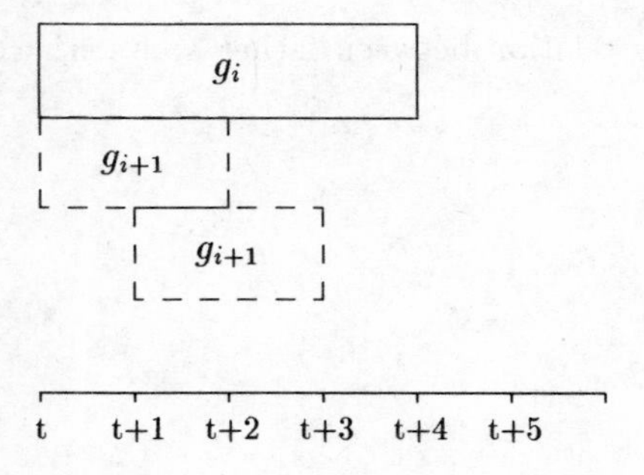

Figure 5.4: Illustration of Interpretation 2

Figures 5.3 and 5.4 illustrate feasible start times of activity g_{i+1} if activity g_i is already scheduled and Interpretation 1 and 2 is used, respectively.

The example in Figure 5.5 shows that the first interpretation will only generate suboptimal solutions, if only the lowest feasible start time assignment is considered. The project consists of four activities, two of which are dummy ones, the start activity 1 and the finish activity 4. Between activity 2 and 3 there are no precedence

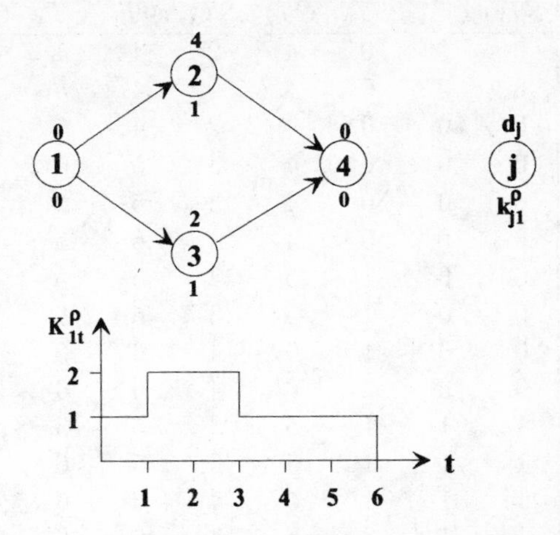

Figure 5.5: Example Problem – Precedence Tree

relations. The resource constraints are induced by one renewable resoure. One unit of this resource is available in periods 1, 4, 5 and 6, whereas in periods 2 and 3 two units can be used. Activity 2 and 3 have a duration of 4 and 2 periods, respectively. Every period they are in process they use one unit of the renewable resource. Obviously the optimal solution, considering the makespan criterion, leasts four periods. The start times for activities 1 to 4 are given by $ST_1 = 0$, $ST_2 = 0$, $ST_3 = 1$ and $ST_4 = 4$.

Remark 5.2

Note, if a job j has a start time ST_j and a completion time of $CT_j = ST_j + d_j$ then it uses renewable resources in the periods t, $t = ST_j + 1, \ldots, CT_j$.

We assume an upper bound on the makespan of six periods. If we employ the precedence tree to guide the search through the solution space, there are two branches to be evaluated, i.e. the branches 1-2-3-4 and 1-3-2-4. Performing a complete enumeration by the use of Interpretation 1 and the neglection of resource requirements we yield the precedence feasible start time assignments displayed in Table 5.1 and 5.2.

Branch	ST_1	CT_1	ST_2	CT_2	ST_3	CT_3	ST_4	CT_4	feas.	
$1-2-3-4$	0	0	0	4	2	4	4	4	$-$	
	0	0	0	4	2	4	5	5	$-$	
	0	0	0	4	2	4	6	6	$-$	
	0	0	0	4	3	5	5	5	$-$	
	0	0	0	4	3	5	6	6	$-$	
	0	0	0	4	4	6	6	6	$+$	*
	0	0	1	5	3	5	5	5	$-$	
	0	0	1	5	3	5	6	6	$-$	
	0	0	1	5	4	6	6	6	$-$	
	0	0	2	6	4	6	6	6	$-$	
	1	1	1	5	3	5	5	5	$-$	
	1	1	1	5	3	5	6	6	$-$	
	1	1	1	5	4	6	6	6	$-$	
	1	1	2	6	4	6	6	6	$-$	
	2	2	2	6	4	6	6	6	$-$	

Table 5.1: Precedence Feasible Start Time Assignments Example Problem (a)

The resource feasible assignments are marked with (+), the infeasible ones with (–). If only the lowest resource feasible start time assignments with respect to the first interpretation, marked with (*), are examined then the optimal solution with makespan of four periods is excluded from consideration.

Remark 5.3

The point is, if only the lowest feasible start times are considered (cf. [88], p.15) then Interpretation 2 has to be used instead of Interpretation 1. The correctness of this statement is proven in Theorem 5.1.

Branch	ST_1	CT_1	ST_2	CT_2	ST_3	CT_3	ST_4	CT_4	feas.	
$1-3-2-4$	0	0	0	4	0	2	4	4	$-$	
	0	0	0	4	0	2	5	5	$-$	
	0	0	0	4	0	2	6	6	$-$	
	0	0	1	5	0	2	5	5	$+$	*
	0	0	1	5	0	2	6	6	$+$	
	0	0	2	6	0	2	6	6	$+$	
	0	0	0	4	1	3	4	4	$+$	
	0	0	0	4	1	3	5	5	$+$	
	0	0	0	4	1	3	6	6	$+$	
	0	0	1	5	1	3	5	5	$+$	
	0	0	1	5	1	3	6	6	$+$	
	0	0	2	6	1	3	6	6	$+$	
	0	0	0	4	2	4	4	4	$-$	
	0	0	0	4	2	4	5	5	$-$	
	0	0	0	4	2	4	6	6	$-$	
	0	0	1	5	2	4	5	5	$-$	
	0	0	1	5	2	4	6	6	$-$	
	0	0	2	6	2	4	6	6	$-$	
	0	0	1	5	3	5	5	5	$-$	
	0	0	1	5	3	5	6	6	$-$	
	0	0	2	6	3	5	6	6	$-$	
	0	0	2	6	4	6	6	6	$-$	
	1	1	1	5	1	3	5	5	$+$	
	1	1	1	5	1	3	6	6	$+$	
	1	1	2	6	1	3	6	6	$+$	
	1	1	1	5	2	4	5	5	$-$	
	1	1	1	5	2	4	6	6	$-$	
	1	1	2	6	2	4	6	6	$-$	
	1	1	1	5	3	5	5	5	$-$	
	1	1	1	5	3	5	6	6	$-$	
	1	1	2	6	3	5	6	6	$-$	
	1	1	2	6	4	6	6	6	$-$	
	2	2	2	6	2	4	6	6	$-$	
	2	2	2	6	3	5	6	6	$-$	
	2	2	2	6	4	6	6	6	$-$	

Table 5.2: Precedence Feasible Start Time Assignments Example Problem (b)

We extend our considerations to the multi-mode case. For notational convenience we need some definitions.

Definition 5.1

(a) An i-partial schedule $\mathcal{PS}_i$ is an i-tuple of quadruples

$$\mathcal{PS}_i = \begin{pmatrix} 1 & , & 2 & , \cdots, & i \\ g_1 & , & g_2 & , \cdots, & g_i \\ m_{g_1} & , & m_{g_2} & , \cdots, & m_{g_i} \\ ST_{g_1} & , & ST_{g_2} & , \cdots, & ST_{g_i} \end{pmatrix},$$

where $(j, g_j, m_{g_j}, ST_{g_j})$ outlines that job g_j is scheduled on stage j in mode m_{g_j} with start time ST_{g_j}.

(b) An i-partial schedule $\mathcal{PS}_i$, where all the activities are scheduled, i.e. $i = J$, is called schedule.

(c) Let $\mathcal{PS}_i$ be an i-partial schedule. A schedule

$$\overline{\mathcal{PS}_J} = \begin{pmatrix} 1 & , \cdots, & J \\ \overline{g}_1 & , \cdots, & \overline{g}_J \\ \overline{m_{\overline{g}_1}} & , \cdots, & \overline{m_{\overline{g}_J}} \\ \overline{ST_{\overline{g}_1}} & , \cdots, & \overline{ST_{\overline{g}_J}} \end{pmatrix}$$

with $g_k = \overline{g}_k$, $m_{g_k} = \overline{m_{\overline{g}_k}}$, $ST_{g_k} = \overline{ST_{\overline{g}_k}}$, $k = 1, \ldots, i$, is called completion of the i-partial schedule $\mathcal{PS}_i$.

Definition 5.2

Let $\overline{T}$ denote an upper bound on the projects makespan and $\mathcal{PS}_i$ be an i-partial schedule with corresponding set of currently scheduled activities

$$\mathcal{AS} := \bigcup_{j=1}^{i} \{g_j\}. \tag{5.5}$$

(a) $\mathcal{PS}_i$ *is called precedence feasible, if*

$$ST_{g_h} + d_{g_h m_{g_h}} \leq ST_{g_k} \qquad g_h, g_k \in \mathcal{AS} \text{ with } g_h \in \mathcal{P}_{g_k}. \tag{5.6}$$

(b) $\mathcal{PS}_i$ *is called feasible with respect to the nonrenewable resources, if*

$$\sum_{j=1}^{i} k^{\nu}_{g_j m_{g_j} r} \leq K^{\nu}_r \qquad r \in N. \tag{5.7}$$

(c) $\mathcal{PS}_i$ *is called feasible with respect to the renewable resources, if*

$$\sum_{\substack{j=1 \\ ST_{g_j}+1 \leq t \leq ST_{g_j}+d_{g_j m_{g_j}}}}^{i} k^{\rho}_{g_j m_{g_j} r} \leq K^{\rho}_{rt} \qquad r \in R,\ t = 1, \ldots, \overline{T}. \tag{5.8}$$

(d) $\mathcal{PS}_i$ *is called feasible, if it is precedence feasible and feasible with respect to the renewable and the nonrenewable resources.*

(e) $K^{\nu}_r(\mathcal{PS}_i) := K^{\nu}_r - \sum_{j=1}^{i} k^{\nu}_{g_j m_{g_j} r}$ *(5.9)*
is called leftover capacity of the nonrenewable resource r, $r \in N$, *with respect to the* i*-partial schedule* $\mathcal{PS}_i$.

(f) $K^{\rho}_{rt}(\mathcal{PS}_i) := K^{\rho}_{rt} - \sum_{\substack{j=1 \\ ST_{g_j}+1 \leq t \leq ST_{g_j}+d_{g_j m_{g_j}}}}^{i} k^{\rho}_{g_j m_{g_j} r}$ *(5.10)*
is called leftover capacity of the renewable resource r, $r \in R$, *in period* t, $t = 1, \ldots, \overline{T}$, *with respect to the* i*-partial schedule* $\mathcal{PS}_i$.

Clearly, for a given (complete) schedule the related objective function value can be easily calculated. We say, a feasible i-partial schedule $\mathcal{PS}_i$ is dominated by a feasible k-partial $\overline{\mathcal{PS}}_k$, if for any feasible completion $\mathcal{PS}_J$ of $\mathcal{PS}_i$ the objective function value cannot be better than the objective function value of the best feasible completion of $\overline{\mathcal{PS}}_k$.

We can now state the theorem, which, rouphly speaking, reduces the examination of one path from the root to a leave of the precedence tree to at most one start time assignment for each activity.

Theorem 5.1

Let (P) be a scheduling problem of the type GRCPSP with a regular measure of performance as objective function. If (P) is feasible, then there exists a permutation of the activities $1, \ldots, J$, *denoted as* $g_1, \ldots, g_J$, *and accompanying modes* $m_{g_1}, \ldots, m_{g_J}$, *such that the schedule with start times* $ST_{g_1}, \ldots, ST_{g_J}$ *fulfilling*

(a) $ST_{g_i} \leq ST_{g_{i+1}}, \qquad i = 1, \ldots, J-1,$

(b) ST_{g_i}, $i = 1, \ldots, J$, *is the lowest feasible start time of activity* g_i *in mode* m_{g_i} *with respect to the precedence relations and the leftover capacities of the (i-1)-partial schedule* $\mathcal{PS}_{i-1}$ *and (a)*[3]

represents an optimal solution.

Proof: Since the proof is more technical we will only sketch it out. Due to the feasibility of the problem we can choose an optimal solution with start times $ST_1^*, \ldots, ST_J^*$ and modes $m_1^*, \ldots, m_J^*$. We then order the jobs with respect to nondecreasing start times and define the corresponding permutation. If the permutation fulfills the requirements then we are finished. Otherwise we choose the lowest index i of g_i, $i = 1, \ldots, J$, which vioalates (b). Let k denote this index. Due to the constant per-period request we can shift activity g_k to the left, i.e. we can start it earlier, as far as (a) is met and the newly derived start time fulfills (b) for $j = 1, \ldots, k$. Since the objective function is a regular measure of performance the objective function value does not increase. Repetively applying the procedure leads to an optimal solution which fulfills the requirements. □

5.2 Minimizing the Projects Makespan

In the previous section we described the use of the precedence tree within an enumeration scheme. We will now give a more formal description which will enable us

[3] Note, if $i = 1$ then the leftover capacities are given by K_{rt}^{ρ}, $r \in R$, $t = 1, \ldots, \overline{T}$, and K_r^{ν}, $r \in N$, respectively.

to discuss the algorithm in more detail. Especially the representation of the bounding rules to be presented in Section 5.5 will be substantially simplified. To avoid notational overhead we are leaving priority rules for later discussion and refer to only if there is some degree of freedom within the algorithm. The notation used to describe the algorithm is displayed in Table 5.3. The algorithm itself is presented in Table 5.4.

To avoid the case distinction whether any activity is already scheduled or not we define $g_0 := 0$, $ST_{g_0} := 0$, $CT_{g_0} := 0$ and $\mathcal{P}_1 := \{0\}$. The set of eligible activities on the first stage ($i = 1$) is defined as $Y_1 := \{1\}$ with corresponding number of eligible activities within the first stage $\hat{N}_1 := |Y_1| = 1$. The first and only activity to be tested is the unique start activity, thus we have $g_1 = Y_{11} = 1$. In general we denote with Y_{iN_i} the N_i'th element of the eligible set Y_i.

Remark 5.4

Obviously the assignment $g_i := Y_{iN_i}$ assumes an ordering of the set of eligible activities on stage i. This ordering can be induced by the job number or other numerical criteria, like e.g. the latest finish time LF_j of activity j derived from critical path analysis. The orderings are referred to as priority rules and will be dealt with in Section 5.4. If not otherwise mentioned, we assume Y_i to be ordered with respect to increasing job number.

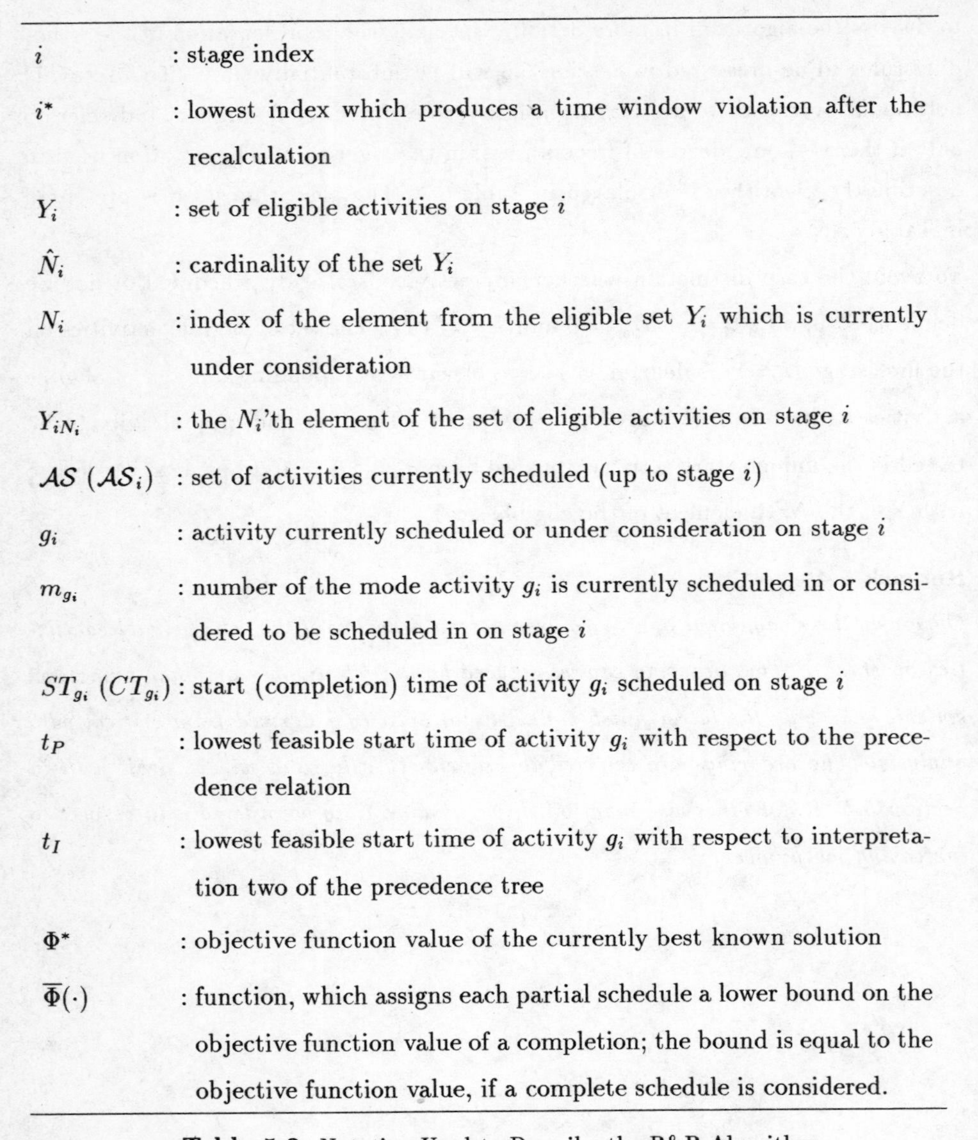

i	: stage index
i^*	: lowest index which produces a time window violation after the recalculation
Y_i	: set of eligible activities on stage i
$\hat{N}_i$	: cardinality of the set Y_i
N_i	: index of the element from the eligible set Y_i which is currently under consideration
Y_{iN_i}	: the N_i'th element of the set of eligible activities on stage i
$\mathcal{AS}$ ($\mathcal{AS}_i$)	: set of activities currently scheduled (up to stage i)
g_i	: activity currently scheduled or under consideration on stage i
m_{g_i}	: number of the mode activity g_i is currently scheduled in or considered to be scheduled in on stage i
ST_{g_i} (CT_{g_i})	: start (completion) time of activity g_i scheduled on stage i
t_P	: lowest feasible start time of activity g_i with respect to the precedence relation
t_I	: lowest feasible start time of activity g_i with respect to interpretation two of the precedence tree
Φ^*	: objective function value of the currently best known solution
$\overline{\Phi}(\cdot)$	: function, which assigns each partial schedule a lower bound on the objective function value of a completion; the bound is equal to the objective function value, if a complete schedule is considered.

Table 5.3: Notation Used to Describe the B&B-Algorithm

Step 1: (Initialization)

$\mathcal{AS} := \emptyset$; $g_0 := 0$; $ST_{g_0} := 0$; $CT_{g_0} := 0$; $\mathcal{P}_1 := \{0\}$; $i := 1$; $Y_1 := \{1\}$; $\hat{N}_1 := 1$; $N_1 := 1$; $g_1 := 1$; $m_1 := 0$;

Step 2: (Select next untested mode or descendant)

If $m_{g_i} < M_{g_i}$ then $m_{g_i} := m_{g_i} + 1$ and goto Step 4;

If $N_i < \hat{N}_i$ then $N_i := N_i + 1$; $g_i := Y_{iN_i}$; $m_{g_i} := 1$ and goto Step 4;

Step 3: (One-stage backtracking)

$i := i - 1$; if $i = 0$ then STOP, else remove job g_i from partial schedule; $\mathcal{AS} := \mathcal{AS} \backslash \{g_i\}$; readjust resource arrays and goto Step 2;

Step 4: (Find a feasible start time)

$t_P := \max\{CT_k; k \in \mathcal{P}_{g_i}\}$; $t_I := ST_{g_{i-1}}$; $t^* := \max\{t_P, t_I\}$; calculate the **earliest** resource feasible start time $\bar{t}$, $t^* \leq \bar{t} \leq LF_{g_i} - d_{g_i m_{g_i}}$, of job g_i in mode m_{g_i}; if scheduling is impossible goto Step 2, else set $ST_{g_i} := \bar{t}$; $CT_{g_i} := \bar{t} + d_{g_i m_{g_i}}$; $\mathcal{AS} := \mathcal{AS} \cup \{g_i\}$ and adjust resource arrays;

Step 5: If i=J then goto Step 7;

Step 6: (Update the eligible set)

$i := i + 1$; calculate the new descendant set $Y_i := Y_{i-1} \backslash \{g_{i-1}\} \cup \{k \in \mathcal{S}_{g_{i-1}}; \mathcal{P}_k \subseteq \mathcal{AS}\}$; set $\hat{N}_i := |Y_i|$; $N_i := 1$; $g_i := Y_{i1}$; $m_{g_i} := 0$ and goto Step 2;

Step 7: (Storing solution and adjusting time bounds)

Store solution g_j, m_{g_j}, ST_{g_j}, $\quad j = 1, \ldots, J$;

Set $LS_j := LS_j - (LF_J - CT_J + 1)$, $\quad j = 1, \ldots, J$;

and $LF_j := LF_j - (LF_J - CT_J + 1)$, $\quad j = 1, \ldots, J$;

Step 8: (Calculating lowest indexed stage violating the time window)

$i^* := \min\{k \in \{1, \ldots, J\}; CT_{g_k} > LF_{g_k}\}$;

Step 9: (Variable-stage backtracking)

Readjust resources used (consumed) by jobs g_k in mode m_{g_k}, $k = J, \ldots, i^*$; $\mathcal{AS} := \mathcal{AS} \backslash \{g_J, \ldots, g_{i^*}\}$; $i := i^*$ and goto Step 2.

Table 5.4: Minimizing the Projects Makespan

In Step 2 the mode and/or descendant is selected for assignment. Two cases have to be considered seperately. First, if there are further modes for the descendant, which is currently examined, that is, $m_{g_i} < M_{g_i}$, then the next mode is selected, i.e. $m_{g_i} := m_{g_i} + 1$. Second, if for the current descendant g_i the last mode has been tested, that is, $m_{g_i} = M_{g_i}$, then the next descendant is choosen and the first mode is selected for assignment, i.e. $N_i := N_i + 1$, $g_i := Y_{iN_i}$, and $m_{g_i} := 1$. If no more modes or descendants are available, then one-stage backtracking (Step 3) has to be performed, otherwise we goto Step 4.

In Step 4 we firstly calculate the lowest feasible start time t_P with respect to the currently scheduled activities and the precedence relations, that is, $t_P := max\{CT_k; k \in \mathcal{P}_{g_i}\}$. Subsequently we calculate the minimal feasible start time t_I due to the second interpretation of the precedence tree and then we set $t^* := max\{t_P, t_I\}$. We then scan the interval $[t^*, LF_{g_i} - d_{g_i m_{g_i}}]$ for the earliest contiguous interval $d_{g_i m_{g_i}}$ periods long where activity g_i can be scheduled without violating the renewable resource constraints. That is, we search $\bar{t}$, $t^* \leq \bar{t} \leq LF_{g_i} - d_{g_i m_{g_i}}$, such that the leftover capacities $K^{\rho}_{rt}(\mathcal{PS}_{i-1})$, $r \in R$, $t = \bar{t}+1, \ldots, \bar{t} + d_{g_i m_{g_i}}$, are greater or equal than the requirements $k^{\rho}_{g_i m_{g_i} r}$. Additionally, the feasibility with respect to nonrenewable resources has to be checked, i.e. $k^{\nu}_{g_i m_{g_i} r} \leq K^{\nu}_{r}(\mathcal{PS}_{i-1})$, $r \in N$, is required.

If feasibility can be assured, we set the start time ST_{g_i} of activity g_i equal to $\bar{t}$ and the completion time CT_{g_i} equal to $ST_{g_i} + d_{g_i m_{g_i}}$. Furthermore, the set of currently scheduled jobs $\mathcal{AS}$, is updated, i.e. $\mathcal{AS} := \mathcal{AS} \cup \{g_i\}$, and the leftover capacities are adjusted. If feasibility cannot be assured, we return to Step 2 and determine the next job/mode combination. Successfully scheduling of activity g_i leads to Step 5.

In Step 5 we check whether all the activities are scheduled. If this holds true, we have found the first feasible solution or an improved solution and we can skip to Step 7, where the new solution is stored and the critical path bounds LS_j, LF_j, $j = 1, \ldots, J$, are recalculated. Since our goal is the improvement of the currently best known solution LS_j and LF_j are reduced by the improvement of the former best known solution incremented by one, that is $LF_J - CT_J + 1$.

If not all the activities are scheduled, we step over to the next stage, $i := i + 1$, and calculate the new descendants, i.e. set of eligible activities Y_i, the number of descendants $\hat{N}_i := |Y_i|$, select the first descendant $N_i := 1$, $g_i := Y_{i1}$, initialize the mode selector variable $m_{g_i} := 0$ and goto Step 2.

After the adaptation of the critical path bounds in Step 7 we determine the lowest stage index i^*, where the newly derived critical path bounds are violated by the current schedule. Variable-stage backtracking is then performed in Step 9, that is, all the activities g_i, which have been scheduled on a stage i, $i \geq i^*$, are removed from the schedule, the resource availabilities are readjusted and the current stage index is set equal to i^*.

The algorithm terminates if in Step 2 no more job/mode assignment is possible and decrementing the stage index in Step 3 leads to $i = 0$. Otherwise, if the stage index is greater than zero after decrementing it in Step 3, then g_i is removed from the partial schedule and resource availabilities are readjusted.

5.3 Optimizing any Regular Measure of Performance

In this section we will adapt the algorithm presented for the investigation of any regular measure of performance. Due to Theorem 5.1 only minor changes are necessary in order to attack scheduling problems of the type GRCPSP with a regular measure of performance as objective function. That is, the enumeration is again guided by the precedence tree and only the evaluation of bounds on the objective function value has to be adapted and explicitly incorporated (Step 4 and Steps 7-9), respectively.

Performing Step 4 as described above follows two intensions. On the one hand, if an activity g_i (in mode m_{g_i}) is assigned a start time $\bar{t}$, $\bar{t} \leq LF_{g_i} - d_{g_i m_{g_i}}$, then the partial schedule $\mathcal{PS}_i$ is completable at least with respect to the precedence relations. On the other hand, since the latest finish times are adjusted after the improvement of the current best solution, the adaptation of the latest start and finish times (Step 7) offers a new bound on the objective function value (makespan) for the completion of the partial schedule. That is, if we would assign a start time $\bar{t}$, $\bar{t} > LF_{g_i} - d_{g_i m_{g_i}}$ to an activity g_i then the partial schedule is not completable with a makespan T, $T \leq LF_J$.

For notational convenience, we denote the objective function with Φ and let Φ^* denote the objective function value of the current best solution. Furthermore, $\overline{\Phi}(\cdot)$ denotes a function which determines a lower bound on the objective function of the completion for a given partial schedule $\mathcal{PS}_i$. With this in mind, we rearrange Step 4 and Steps 7-9 to Step 4' and Step 7' (cf. Table 5.5).

Step 4': (Find a feasible start time)
$t_P := \max\{CT_k; k \in \mathcal{P}_{g_i}\}$; $t_I := ST_{g_{i-1}}$; $t^* := \max\{t_P, t_I\}$; calculate the **earliest** resource feasible start time $\bar{t}$, $t^* \leq \bar{t} \leq LF_{g_i} - d_{g_i m_{g_i}}$, of job g_i in mode m_{g_i}; if scheduling is impossible or $\overline{\Phi}(\mathcal{PS}_i) \geq \Phi^*$ then goto Step 2, else set $ST_{g_i} := \bar{t}$; $CT_{g_i} := \bar{t} + d_{g_i m_{g_i}}$; $\mathcal{AS} := \mathcal{AS} \cup \{g_i\}$ and adjust resource arrays;

Step 7': (Storing solution and adjusting the bounds)
Store solution g_i, m_{g_i}, ST_{g_i}, $i = 1, \ldots, J$, Φ^*;
remove job g_J from partial schedule; $\mathcal{AS} := \mathcal{AS} \backslash \{g_J\}$; readjust resource arrays and goto Step 2;

Table 5.5: Optimizing a Regular Measure of Performance

Remark 5.5

In contrast to [88], pp. 13, we included lower bound evaluation explicitly and prevent computational overhead by restarting the procedure after each determination of a new current best solution.

5.4 Priority Rules and Heuristic Search Strategies

The general strategy of the algorithm presented is as follows: It starts with an upper bound on the objective funtion value. Subsequently the algorithm tries to find a solution with an objective function value which is less than or equal to the known bound, and then seeks to improve it by systematically scanning the search space. Unfortunately, the problem at hand has turned out to be a member of the class of NP-hard problems (cf. [55]). Hence, the probability to find a polynomial algorithm solving the problem is quite low and thus the computational effort to spend on the investigation grows rapidly with the number of jobs. However, the algorithm presented may be stopped if the tradeoff between the improvement of the solution and the computational effort becomes unfavourable (truncated exact methods). Another way to involve heuristic elements in an exact solution procedure is as follows: Whenever there is some kind of freedom in the current decision on the choice of the variable to fix, then employ a criterion to support this decision. The addressed freedom in the algorithm is the decision which eligible activity to schedule first in which mode.

The mode is commonly selected depending on the objective function under consideration. That is, e.g. with respect to non-decreasing duration and non-increasing cashflow if the makespan and the net present value is considered, repectively (cf. [115]). A simple priority rule is to choose the activity with the lowest activity number within the eligible set. Other priority rules are displayed in Table 5.6 (cf. [115]), where d_j^{min}, d_j^{max} and $\overline{d}_j$ denote the minimum, maximum and average duration, respectively, of activity j, $j = 1, \ldots, J$.

Note, all the priority rules can be used within a single pass strategy. In a single

No.	Description	Formula
1.	minimum job number	j
2.	randomly	-
3.	maximum duration	d_j^{max}
4.	maximum average duration	$\overline{d}_j$
5.	minimum latest finish time	LF_j
6.	minimum latest finish time reduced by smallest duration	$LF_j - d_j^{min}$
7.	minimum latest finish time reduced by average duration	$LF_j - \overline{d}_j$
8.	minimum possible start time subject to the precedence relations	$max\{CT_k; k \in \mathcal{P}_j\}$

Table 5.6: Priority Rules

pass strategy at each node of the B&B tree the decision on the activity to schedule is performed exactly once. The job is selected in accordance with the priority rule (cf. [28], [67], [70], [71], [114], [118]) in order to obtain a heuristic solution for the single-mode problem.

In the multi-mode-problem additional efforts have to be undertaken in order to select a mode. Additionally, there is a feasibility problem if nonrenewable resources have to be taken into account.

To overcome the difficulties related with the single-pass strategy Drexl and Grünewald (cf. [41], [43]) developed a weighted random technique. For each job/mode combination $[j, m]$ relating to jobs of the eligible set a positive weight is calculated, e.g.

$$w_{jm} := (max\{d_{kl}; k \in Y_i, l = 1, \ldots, M_k\} - d_{jm} + \epsilon)^{\alpha}, \tag{5.11}$$

where $\epsilon > 0$ has to secure that the weight is positive and $\alpha \geq 0$ transforms the term $(\cdot)$ exponentially. The differences between the weights are diminished ($\alpha < 1$) or enforced ($\alpha > 1$). The weigths are then transformed in order to define the probabilities to select a specific job/mode combination. A multiple pass strategy is obtained by using the procedure several times.

5.5 Acceleration Schemes

In this section we will present some bounding rules which can be easily incorporated into the algorithm proposed. The main problem one encounters within the development of bounding rules is to keep the additional effort for the verification of the assumptions low in order to get the rules efficient. Therefore, the outline is as follows: First, we describe by problem instances the situations in which the rules are applicable. Then the rules are summarized as theorems which will be proven as well. If the implementation itself is not obvious then additional hints are given. If not explicitly mentioned then the rules are applicable for problems of type GRCPSP with any regular measure of performance as objective function.

In this section, if not otherwise mentioned, we use the notion of an i-partial schedule to refer to a feasible i-partial schedule derived by the algorithm presented in Table 5.4 and 5.5, respectively. Note, for a given i-partial schedule $\mathcal{PS}_i$ and a job/mode combination $[g_{i+1}, m_{g_{i+1}}]$ the start time $ST_{g_{i+1}}$ on stage $(i+1)$ is uniquely determined if it exists.

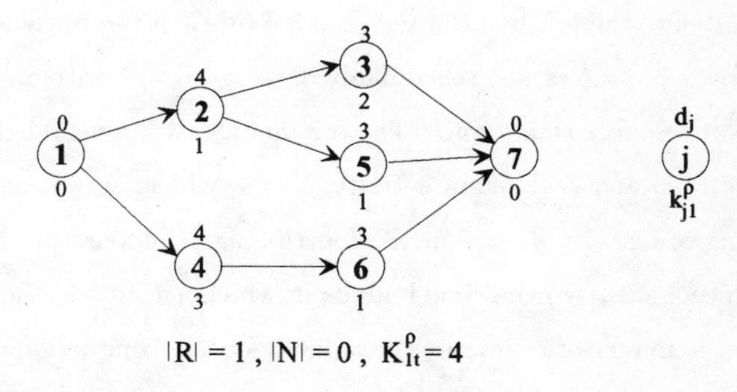

Figure 5.6: Project Instance

We consider the single-mode problem displayed in Figure 5.6 and the corresponding precedence tree (cf. Figure 5.7). We assume an upper bound on the projects

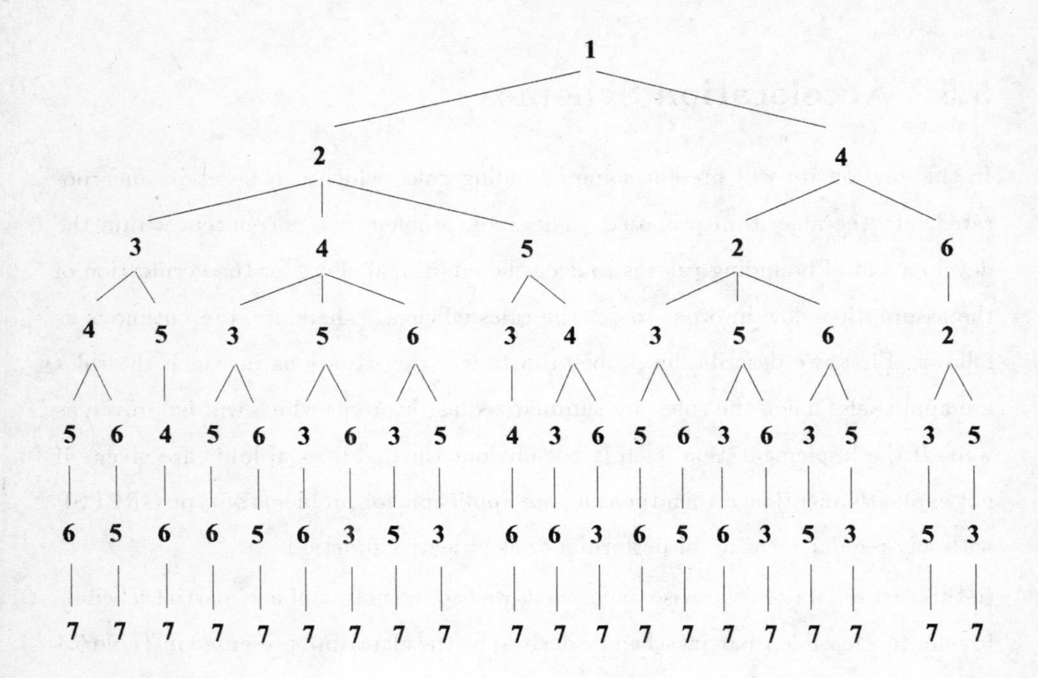

Figure 5.7: Precedence Tree Project Instance

makespan of $\overline{T} = 8$ and an availability of 4 units per period of the only renewable resource. Using the left hand branch 1-2-3-4-5-6-7 of the precedence tree we notice that activity $g_4 = 4$ is not schedulable on stage $i = 4$ without violating the time window- or resource-constraints. The original algorithm would then try to schedule activity $\overline{g}_4 = 5$ and then again activity $\overline{g}_5 = 4$, which, surely, cannot be successful. Scheduling of activity $\overline{g}_4 = 5$ in the fourth stage tightens (due to Interpretation 2 of the precedence tree) the time bounds in which activity 4 is schedulable and, furthermore, it reduces the resources available for the completion of activity 4. Thus after recognizing that an activity is not schedulable backtracking should occur in order to free resources for its completion.

If a multi-mode problem is considered then the fact that a job j is not schedulable on a stage i means that, more precisely, job j is not schedulable in any mode. The rule is summarized in the following theorem:

Theorem 5.2 *(Bounding Rule 1)*
If a job g_{i+1} is not schedulable (w.r.t. precedence-, time window- or resource-constraints) on stage $(i+1)$ with feasible i-partial schedule

$$\mathcal{PS}_i = \begin{pmatrix} 1 & , & 2 & , \cdots , & i \\ g_1 & , & g_2 & , \cdots , & g_i \\ m_{g_1} & , & m_{g_2} & , \cdots , & m_{g_i} \\ ST_{g_1} & , & ST_{g_2} & , \cdots , & ST_{g_i} \end{pmatrix},$$

then it is not schedulable on stage $(i+k+1)$ with $(i+k)$-partial schedule $\mathcal{PS}_{i+k}$

$$\mathcal{PS}_{i+k} = \begin{pmatrix} 1 & , & 2 & , \cdots , & i & , \cdots , & i+k \\ g_1 & , & g_2 & , \cdots , & g_i & , \cdots , & g_{i+k} \\ m_{g_1} & , & m_{g_2} & , \cdots , & m_{g_i} & , \cdots , & m_{g_{i+k}} \\ ST_{g_1} & , & ST_{g_2} & , \cdots , & ST_{g_i} & , \cdots , & ST_{g_{i+k}} \end{pmatrix}.$$

Proof: Obvious. □

Surely, the underlying idea is adaptable for a single job/mode combination too, that is, if a job g_{i+1} is not schedulable in mode $m_{g_{i+1}}$ on stage $(i+1)$, with i-partial schedule $\mathcal{PS}_i$, then $[g_{i+1}, m_{g_{i+1}}]$ is not schedulable on a stage $(i+k+1)$ with $(i+k)$-partial schedule $\mathcal{PS}_{i+k}$. Unfortunately, the effect of this adaption is completely consumed by the additional effort to be performed.

The instance presented turns out another point of attack for the reduction of computational time to spend on solving the problem. If we compare the branch 1-2-4 and the branch 1-4-2 of the precedence tree, we yield the following start time assignments $ST_1 = 0$, $ST_2 = 0$, $ST_4 = 0$ and $\overline{ST}_1 = 0$, $\overline{ST}_2 = 0$, $\overline{ST}_4 = 0$, respectively.

That is, different paths within the precedence tree lead to the same solutions. Therefore the second path should be excluded from evaluation. The fact itself is simply stated in Theorem 5.3 and needs no further explanation.

Theorem 5.3 *(Bounding Rule 2)*

We consider the following feasible $(i+2)$-partial schedules

$$\mathcal{PS}_{i+2} = \begin{pmatrix} 1 & , & 2 & ,\cdots, & i & , & i+1 & , & i+2 \\ g_1 & , & g_2 & ,\cdots, & g_i & , & g_{i+1} & , & g_{i+2} \\ m_{g_1} & , & m_{g_2} & ,\cdots, & m_{g_i} & , & m_{g_{i+1}} & , & m_{g_{i+2}} \\ ST_{g_1} & , & ST_{g_2} & ,\cdots, & ST_{g_i} & , & ST_{g_{i+1}} & , & ST_{g_{i+2}} \end{pmatrix}$$

and

$$\overline{\mathcal{PS}}_{i+2} = \begin{pmatrix} 1 & , & 2 & ,\cdots, & i & , & i+1 & , & i+2 \\ g_1 & , & g_2 & ,\cdots, & g_i & , & \overline{g}_{i+1} & , & \overline{g}_{i+2} \\ m_{g_1} & , & m_{g_2} & ,\cdots, & m_{g_i} & , & \overline{m}_{\overline{g}_{i+1}} & , & \overline{m}_{\overline{g}_{i+2}} \\ ST_{g_1} & , & ST_{g_2} & ,\cdots, & ST_{g_i} & , & \overline{ST}_{\overline{g}_{i+1}} & , & \overline{ST}_{\overline{g}_{i+2}} \end{pmatrix}$$

If we have

$$\left.\begin{array}{lllll} g_{i+1} & = \overline{g}_{i+2} & , g_{i+2} & = & \overline{g}_{i+1} \\ m_{g_{i+1}} & = \overline{m}_{\overline{g}_{i+2}} & , m_{g_{i+2}} & = & \overline{m}_{\overline{g}_{i+1}} \\ ST_{g_{i+1}} & = \overline{ST}_{\overline{g}_{i+2}} & = ST_{g_{i+2}} & = & \overline{ST}_{\overline{g}_{i+1}} \end{array}\right\} \tag{5.12}$$

then the completions of $\overline{\mathcal{PS}}_{i+2}$ are dominated by the completions of $\mathcal{PS}_{i+2}$.

Proof: Obvious. □

The rule itself has already been mentioned in the literature (cf. [41]) for a less general framework. Obviously, it is extendable: We consider an i-partial schedule $\mathcal{PS}_i$ and activities $g_{i+1}, \ldots, g_{i+k}$, with accompanying modes $m_{g_{i+1}}, \ldots, m_{g_{i+k}}$. If we obtain for all the permutations of $g_{i+1}, \ldots, g_{i+k}$, scheduled in the modes $m_{g_{i+1}}, \ldots, m_{g_{i+k}}$, on stages $i+1, \ldots, i+k$, start times $ST_{g_{i+1}} = \ldots = ST_{g_{i+k}}$, which do not depend on the permutation of $g_{i+1}, \ldots, g_{i+k}$, then only one permutation has to be examined. That is, $(k-1)!$ evaluations can be saved. The procedure derived in [41] is "a hybrid Branch and Bound/dynamic programming algorithm, which combines the advantages of both approaches. It should be noted that the implementation of this procedure is computationally highly involved and indeed requires a lot of administrative instructions....". That is, as outlined above, the problem is the verification of (5.12) and the comparison of $\mathcal{PS}_i$ and $\overline{\mathcal{PS}}_i$. In the following we will propose a very efficient test of the assumptions of Theorem 5.3. For checking this, the algorithm's Step 4' is changed to Step 4" (cf. Table 5.7), where the three-dimensional integer-array $PT[i][g][m]$ is initialized with -1.

Step 4": $t_P := \max\{CT_k; k \in \mathcal{P}_{g_i}\}$; $t_I := ST_{g_{i-1}}$; $t^* := \max\{t_P, t_I\}$; calculate the **earliest** resource feasible start time $\bar{t}$, $t^* \le \bar{t} \le LF_{g_i} - d_{g_i m_{g_i}}$, of job g_i in mode m_{g_i}; if scheduling is impossible then goto 2;
$PT[i+1][g_i][m_{g_i}] := \bar{t}$, if $\overline{\Phi}(\mathcal{PS}_{i+1}) \ge \Phi^*$ then goto 2, else set $ST_{g_i} := \bar{t}$;
$CT_{g_i} := \bar{t} + d_{g_i m_{g_i}}$; $\mathcal{AS} := \mathcal{AS} \cup \{g_i\}$ and adjust resource arrays;

Table 5.7: Extension of the Algorithm by Bounding Rule 2

The following theorem offers a criterion to check the assumptions of the bounding rule efficiently.

Theorem 5.4

We assume $d_{jm} > 0$, $j = 2, \ldots, J-1$, $m = 1, \ldots, M_j$. Let $i \ge 1$ and $\overline{\mathcal{PS}}_{i+2}$ be a feasible $(i+2)$-partial schedule derived by the modified algorithm. If $\overline{ST}_{\overline{g}_{i+1}} = \overline{ST}_{\overline{g}_{i+2}}$, $N_{i+1} > N_{i+2}$ and $PT[i+2][\overline{g}_{i+2}][\overline{m}_{\overline{g}_{i+2}}] = \overline{ST}_{\overline{g}_{i+2}}$ then the assumptions of Theorem 5.3 are fulfilled, that is, the completions of $\overline{\mathcal{PS}}_{i+2}$ are dominated by *previously evaluated completions of $\overline{\mathcal{PS}}_i$.*

Proof: We assumed $d_{jm} > 0$, $j = 2, \dots, J-1$, $m = 1, \dots, M_j$, therefore the equality $\overline{ST}_{\overline{g}_{i+1}} = \overline{ST}_{\overline{g}_{i+2}}$ implies $\overline{g}_{i+1}, \overline{g}_{i+2} \in Y_{i+1}$.

Since $\overline{g}_{i+2}$ is schedulable on stage $i+2$ in mode $\overline{m}_{\overline{g}_{i+2}}$ with $(i+1)$-partial schedule

$$\overline{\mathcal{PS}}_{i+1} = \begin{pmatrix} 1 & , & 2 & , \cdots, & i & , & i+1 \\ g_1 & , & g_2 & , \cdots, & g_i & , & \overline{g}_{i+1} \\ m_{g_1} & , & m_{g_2} & , \cdots, & m_{g_i} & , & \overline{m}_{\overline{g}_{i+1}} \\ ST_{g_1} & , & ST_{g_2} & , \cdots, & ST_{g_i} & , & \overline{ST}_{\overline{g}_{i+1}} \end{pmatrix}$$

and start time $\overline{ST}_{\overline{g}_{i+2}}$, it is also schedulable on stage $(i+1)$ with i-partial schedule $\overline{\mathcal{PS}}_i$.

Let $Y_{i+1} = \{h_1, \dots, h_{\hat{N}_{i+1}}\}$ be the (ordered) eligible set of stage $(i+1)$ corresponding to the i-partial schedule $\overline{\mathcal{PS}}_i$ with $h_k = \overline{g}_{i+1}$, i.e. $k = N_{i+1}$.

Since the network is assumed to be numerically labeled and furthermore, the eligible set is ordered with respect to increasing job numbers we now have $Y_{i+2} = \{h_1, \dots, h_{k-1}, l_k, \dots, l_{\hat{N}_{i+2}}\}$ with $\overline{g}_{i+2} = h_l$, i.e. $l = N_{i+2} \le k-1 < k = N_{i+1}$. Thus $PT[i+2][\overline{g}_{i+2}][\overline{m}_{\overline{g}_{i+2}}]$ is the start time of $\overline{g}_{i+2}$ in mode $\overline{m}_{\overline{g}_{i+2}}$ scheduled on stage $(i+1)$ with i-partial schedule $\overline{\mathcal{PS}}_i$. Using $\overline{ST}_{\overline{g}_{i+1}} = \overline{ST}_{\overline{g}_{i+2}} = PT[i+2][g_{i+2}][m_{g_{i+2}}]$ we conclude $ST_{\overline{g}_{i+1}} = \overline{ST}_{\overline{g}_{i+2}}$ in the $(i+2)$-partial schedule

$$\mathcal{PS}_{i+2} = \begin{pmatrix} 1 & , & 2 & , \cdots, & i & , & i+1 & , & i+2 \\ g_1 & , & g_2 & , \cdots, & g_i & , & \overline{g}_{i+2} & , & \overline{g}_{i+1} \\ m_{g_1} & , & m_{g_2} & , \cdots, & m_{g_i} & , & \overline{m}_{\overline{g}_{i+2}} & , & \overline{m}_{\overline{g}_{i+1}} \\ ST_{g_1} & , & ST_{g_2} & , \cdots, & ST_{g_i} & , & \overline{ST}_{\overline{g}_{i+2}} & , & ST_{\overline{g}_{i+1}} \end{pmatrix}.$$

Since $N_{i+1} > N_{i+2}$ the completions of $\mathcal{PS}_{i+2}$ have been previously examined and the theorem is proven. □

Remark 5.6

We assume the eligible set $Y_i = \{h_1, \ldots, h_{\hat{N}_i}\}$ *is ordered with respect to non-decreasing priority values* $\Pi(h)$*, i.e.* $\Pi(h_i) \leq \Pi(h_{i+1})$, $i = 1, \ldots, \hat{N}_i - 1$. *If* $Y_{i+1} = \{h_1, \ldots, h_{\hat{N}_i}\} \backslash \{h_l\} \cup \{h \in \mathcal{S}_{h_l}; \mathcal{P}_h \subset \mathcal{AS}\}$ *can be ordered by priority-rule* Π*, such that* $\Pi(h_r) \leq \Pi(g)$ *for all* $g \in \{h \in \mathcal{S}_{h_l}; \mathcal{P}_h \subset \mathcal{AS}\}$ *and* $r < l$ *then Theorem 5.4 is also applicable with the priority rule* $\Pi(\cdot)$.

In order to illustrate the effect of the rule if more than two jobs have the same start time, we make use of the example displayed in Figure 5.1. We assume that the activities 2, 3 and 4 are schedulable with the same start time. That is, if scheduling of the activities 2, 3 and 4 leads to the start times $ST_2 = ST_3 = ST_4 = 0$ independently of the sequence the activities are scheduled in, then applying bounding rule 2 prunes the precedence tree as depicted in Figure 5.8, where (—) denotes the pruning of a branch.

The following bounding rule makes use of the fact that the set of semi-active schedules is a dominant set with respect to any regular measure of performance (cf. Chapter 4). That is, for any regular measure of perfomance there is an optimal schedule which is semi-active. The bounding rule is the so called left shift rule.

We use the example given in Figure 5.6, where we redefine $d_4 = 1$ and $k_{41}^{\rho} = 2$. The partial schedules derived from the branch (1-2-3-4) of the precedence tree are dominated by partial schedules derived from the branch (1-2-4-3) (cf. Figure 5.9). Note, the schedules we obtain by using the branch (1-2-3-4) are not semi-active, since activity 4 can be locally left shifted. The rule has already been mentioned and sucessfully implemented for the single-mode case by e.g. Demeulemeester (cf. [32]). It is stated in the following theorem and the proof is given for the sake of completeness.

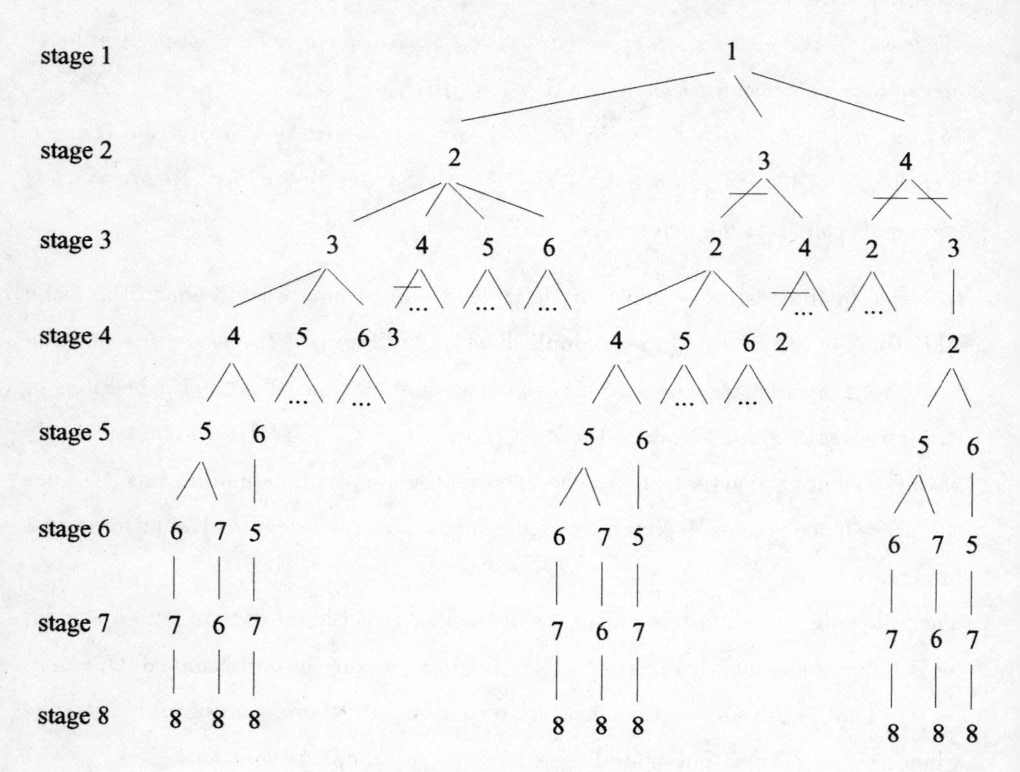

Figure 5.8: Pruned Precedence Tree

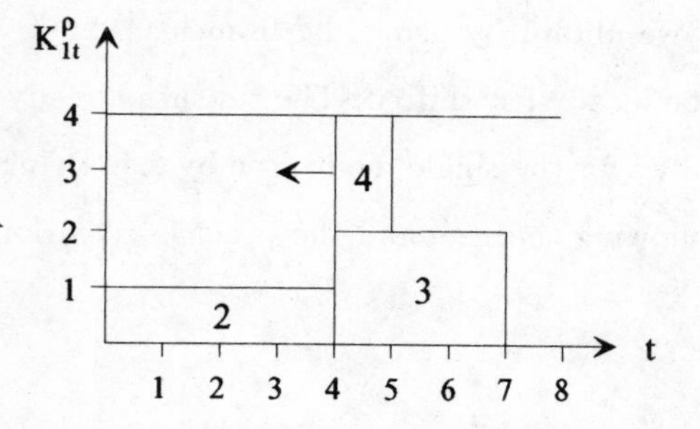

Figure 5.9: Left-Shift-Rule

Theorem 5.5 *(Bounding Rule 3, Left-Shift-Rule)*

We consider the following feasible i-partial schedule

$$\mathcal{PS}_i = \begin{pmatrix} 1 & , & 2 & ,\cdots, & i \\ g_1 & , & g_2 & ,\cdots, & g_i \\ m_{g_1} & , & m_{g_2} & ,\cdots, & m_{g_i} \\ ST_{g_1} & , & ST_{g_2} & ,\cdots, & ST_{g_i} \end{pmatrix}.$$

Let $ST_{g_{i+1}}$ denote the start time of activity g_{i+1} in mode $m_{g_{i+1}}$ on stage $i+1$ with partial schedule $\mathcal{PS}_i$. If activity g_{i+1} is additionally schedulable in mode $m_{g_{i+1}}$ with a start time $\overline{ST}_{g_{i+1}}$, $\overline{ST}_{g_{i+1}} < ST_{g_i}$, without violating the precedence- and (renewable) resource-constraints, then the $(i+1)$-partial schedule

$$\mathcal{PS}_{i+1} = \begin{pmatrix} 1 & , & 2 & ,\cdots, & i & , & i+1 \\ g_1 & , & g_2 & ,\cdots, & g_i & , & g_{i+1} \\ m_{g_1} & , & m_{g_2} & ,\cdots, & m_{g_i} & , & m_{g_{i+1}} \\ ST_{g_1} & , & ST_{g_2} & ,\cdots, & ST_{g_i} & , & ST_{g_{i+1}} \end{pmatrix}$$

does not have to be examined.

Proof: Let $k := \min\{l \leq i; ST_{g_l} = ST_{g_i}\}$. Since $\mathcal{PS}_{i+1}$ is feasible the $(i+1)$-partial schedule

$$\overline{\mathcal{PS}}_{i+1} := \begin{pmatrix} 1 & ,\cdots, & k-1 & , & k & , & k+1 & ,\cdots, & i+1 \\ g_1 & ,\cdots, & g_{k-1} & , & g_{i+1} & , & g_k & ,\cdots, & g_i \\ m_{g_1} & ,\cdots, & m_{g_{k-1}} & , & m_{g_{i+1}} & , & m_{g_k} & ,\cdots, & m_{g_i} \\ ST_{g_1} & ,\cdots, & ST_{g_{k-1}} & , & \overline{ST}_{g_{i+1}} & , & ST_{g_k} & ,\cdots, & ST_{g_i} \end{pmatrix}$$

is a feasible one, too. Due to the constant per period usage we have

$$K^{\rho}_{rt}(\mathcal{PS}_{i+1}) \leq K^{\rho}_{rt}(\overline{\mathcal{PS}}_{i+1}) \qquad r \in R,\ t = ST_{g_{i+1}} + 1, \ldots, \overline{T} \tag{5.13}$$

which proves the theorem. □

Whereas the first three bounding rules offered criteria for the estimation of the objective function value of several completions of partial schedules, the one to be presented next checks feasibility of the completion by underestimating the consumption of nonrenewable resources of a completion. It is designed for the cases $|D| > 0$ and/or $|N| > 0$.

Theorem 5.6 *(Bounding Rule 4)*
We consider the following feasible i-partial schedule

$$\mathcal{PS}_i = \begin{pmatrix} 1 & , & 2 & ,\cdots, & i \\ g_1 & , & g_2 & ,\cdots, & g_i \\ m_{g_1} & , & m_{g_2} & ,\cdots, & m_{g_i} \\ ST_{g_1} & , & ST_{g_2} & ,\cdots, & ST_{g_i} \end{pmatrix}$$

and the leftover capacities of nonrenewable resources $K_r^\nu(\mathcal{PS}_i)$, $r \in N$. Let $\overline{\mathcal{AS}} = \{1,\ldots,J\}\backslash\{g_1,\ldots,g_i\}$ be the set of currently unscheduled activities. If there is a resource r, $r \in N$, with

$$K_r^\nu(\mathcal{PS}_i) < \sum_{j\in\overline{\mathcal{AS}}} \min\{k_{jmr}^\nu; m = 1,\ldots,M_j\}$$

then the schedule is not completable.

Proof: Obvious. □

Since the rule checks completability of the current partial schedule it is applicable for any objective function under consideration.

Drexl (cf. [41]) applied the rule in an algorithm for dealing with models suitable for audit staff scheduling. Preliminary tests have successfully been performed. But, employing the following remark the effect of the rule can be substantially improved.

Remark 5.7

The bounding rule of Theorem 5.6 can be easily implemented via preprocessing by calculating:

$$kmin_{jr}^\nu := \min\{k_{jmr}^\nu; m = 1,\ldots,M_j\}, \qquad j = 1,\ldots,J, r \in N, \tag{5.14}$$

and adjusting the input data as follows:

$$\overline{k}^{\nu}_{jmr} := k^{\nu}_{jmr} - kmin^{\nu}_{jr}, \qquad j = 1, \ldots, J, m = 1, \ldots, M_j, r \in N \qquad (5.15)$$

and

$$\overline{K}^{\nu}_{r} := K^{\nu}_{r} - \sum_{j=1}^{J} kmin^{\nu}_{jr}, \qquad r \in N. \qquad (5.16)$$

Remark 5.8

If we introduce for each renewable resource r, $r \in R$, a nonrenewable resource, that is we yield $N' = N \cup R$, where the overall availability is

$$K^{\nu}_{r} := \sum_{t=1}^{\overline{T}} K^{\rho}_{rt},$$

and determine the requests analogously, then the rule is applicable on basis of the overall usage of renewable resources.

The consequence of Theorem 5.6 and Remark 5.7 is that for each nonrenewable resource r, $r \in N$, and each j, $j = 1, \ldots, J$, at least one mode m becomes a consumption of k^{ν}_{jmr} equal to zero, which reduces the resource factor (cf. Section 6.4) of the problem; thus beside the effect of the rule the problems are easier to solve (cf. [68]).

The last rule is especially designed for the multi-mode problem, where only renewable resources have to be taken into consideration.

Theorem 5.7 *(Bounding Rule 5, $|N| = |D| = 0$)*

We consider the following feasible $(i+1)$-partial schedule

$$\mathcal{PS}_{i+1} = \begin{pmatrix} 1 & , 2 & , \cdots, & i & , i+1 \\ g_1 & , g_2 & , \cdots, & g_i & , g_{i+1} \\ m_{g_1} & , m_{g_2} & , \cdots, & m_{g_i} & , m_{g_{i+1}} \\ ST_{g_1} & , ST_{g_2} & , \cdots, & ST_{g_i} & , ST_{g_{i+1}} \end{pmatrix}.$$

Let $CT_{g_{i+1}}^{min}$ *:=minimum of the completion times of job* g_{i+1} *scheduled on stage* $i+1$ *within the modes* $1, \ldots, m_{g_{i+1}}$ *with i-partial schedule*

$$\mathcal{PS}_i = \begin{pmatrix} 1 & , & 2 & , \cdots , & i \\ g_1 & , & g_2 & , \cdots , & g_i \\ m_{g_1} & , & m_{g_2} & , \cdots , & m_{g_i} \\ ST_{g_1} & , & ST_{g_2} & , \cdots , & ST_{g_i} \end{pmatrix}$$

and accompanying mode $m_{g_{i+1}}^{min}$. *Furthermore let* $CT_{g_{i+1}}^{oldmin}$ *denote the minimal completion time of job* g_{i+1} *scheduled on stage* $(i+1)$ *within the modes* $1, \ldots, m_{g_{i+1}}^{min} - 1$ *with i-partial schedule* $\mathcal{PS}_i$. *If the corresponding mode sets are empty, then* $CT_{g_{i+1}}^{min}$ *and* $CT_{g_{i+1}}^{oldmin}$*, respectively, are set to infinity. Then: The completions of a schedule*

$$\overline{\mathcal{PS}}_{i+2} = \begin{pmatrix} 1 & , & 2 & , \cdots , & i & , & i+1 & , & i+2 \\ g_1 & , & g_2 & , \cdots , & g_i & , & g_{i+1} & , & g_{i+2} \\ m_{g_1} & , & m_{g_2} & , \cdots , & m_{g_i} & , & \overline{m}_{g_{i+1}} & , & m_{g_{i+2}} \\ ST_{g_1} & , & ST_{g_2} & , \cdots , & ST_{g_i} & , & \overline{ST}_{g_{i+1}} & , & \overline{ST}_{g_{i+2}} \end{pmatrix}$$

with

$$\text{case (a) } CT_{g_{i+1}}^{min} \leq \overline{ST}_{g_{i+2}} \quad \text{and} \quad \overline{m}_{g_{i+1}} > m_{g_{i+1}}^{min} \tag{5.17}$$

or

$$\text{case (b) } CT_{g_{i+1}}^{oldmin} \leq \overline{ST}_{g_{i+2}} \quad \text{and} \quad \overline{m}_{g_{i+1}} = m_{g_{i+1}}^{min} \tag{5.18}$$

do not have to be examined.

Proof: Let $ST_{g_{i+1}}^{min}$ and $ST_{g_{i+1}}^{oldmin}$ denote the start times corresponding to $CT_{g_{i+1}}^{min}$ and $CT_{g_{i+1}}^{oldmin}$, respectively.

Case (a): We consider the left-over capacities of the $(i+2)$-partial schedule

$$\mathcal{PS}_{i+2} = \begin{pmatrix} 1 & , & 2 & , \cdots, & i & , & i+1 & , & i+2 \\ g_1 & , & g_2 & , \cdots, & g_i & , & g_{i+1} & , & g_{i+2} \\ m_{g_1} & , & m_{g_2} & , \cdots, & m_{g_i} & , & m_{g_{i+1}}^{min} & , & m_{g_{i+2}} \\ ST_{g_1} & , & ST_{g_2} & , \cdots, & ST_{g_i} & , & ST_{g_{i+1}}^{min} & , & ST_{g_{i+2}} \end{pmatrix}$$

and the left-over capacities of the $(i+2)$-partial schedule

$$\overline{\mathcal{PS}}_{i+2} = \begin{pmatrix} 1 & , & 2 & , \cdots, & i & , & i+1 & , & i+2 \\ g_1 & , & g_2 & , \cdots, & g_i & , & g_{i+1} & , & g_{i+2} \\ m_{g_1} & , & m_{g_2} & , \cdots, & m_{g_i} & , & \overline{m}_{g_{i+1}} & , & m_{g_{i+2}} \\ ST_{g_1} & , & ST_{g_2} & , \cdots, & ST_{g_i} & , & \overline{ST}_{g_{i+1}} & , & \overline{ST}_{g_{i+2}} \end{pmatrix}.$$

An illustration of the situation is given in Figure 5.10. Since $\overline{ST}_{g_{i+2}} \geq CT_{g_{i+1}}^{min}$, $\overline{ST}_{g_{i+2}}$

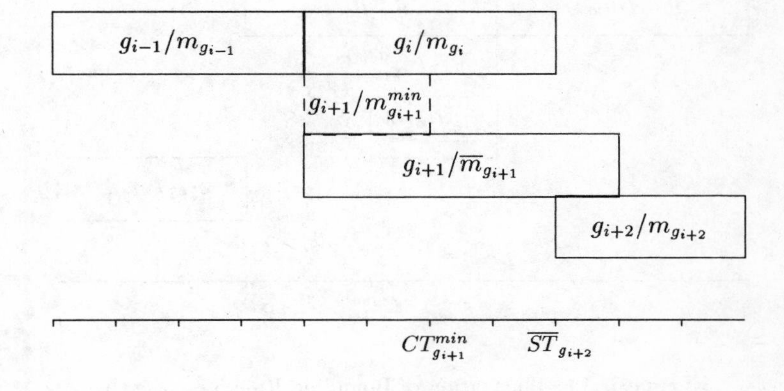

Figure 5.10: Illustration of Bounding Rule 5 – Case (a)

is a feasible start time of activity g_{i+2} in mode $m_{g_{i+2}}$ on stage $(i+2)$ with $(i+1)$-partial schedule $\mathcal{PS}_{i+1}$. Therefore, $ST_{g_{i+2}} \leq \overline{ST}_{g_{i+2}}$, which leads to $K_{rt}^{\rho}(\overline{\mathcal{PS}}_{i+2}) \leq K_{rt}^{\rho}(\mathcal{PS}_{i+2})$ for each period t, $t = \overline{ST}_{g_{i+2}} + 1, \ldots, \overline{T}$, and each renewable resource r, $r \in R$, which proves case (a).

Case (b): Let $m_{g_{i+1}}^{oldmin}$ be the mode accompanying $CT_{g_{i+1}}^{oldmin}$. We consider the left-over capacities of the $(i+2)$-partial schedule

$$\mathcal{PS}_{i+2} = \begin{pmatrix} 1 & , & 2 & , \cdots, & i & , & i+1 & , & i+2 \\ g_1 & , & g_2 & , \cdots, & g_i & , & g_{i+1} & , & g_{i+2} \\ m_{g_1} & , & m_{g_2} & , \cdots, & m_{g_i} & , & m_{g_{i+1}}^{oldmin} & , & m_{g_{i+2}} \\ ST_{g_1} & , & ST_{g_2} & , \cdots, & ST_{g_i} & , & ST_{g_{i+1}}^{oldmin} & , & ST_{g_{i+2}} \end{pmatrix}$$

and the left-over capacities of the $(i+2)$-partial schedule

$$\overline{\mathcal{PS}}_{i+2} = \begin{pmatrix} 1 & , & 2 & , \cdots, & i & , & i+1 & , & i+2 \\ g_1 & , & g_2 & , \cdots, & g_i & , & g_{i+1} & , & g_{i+2} \\ m_{g_1} & , & m_{g_2} & , \cdots, & m_{g_i} & , & m_{g_{i+1}}^{min} & , & m_{g_{i+2}} \\ ST_{g_1} & , & ST_{g_2} & , \cdots, & ST_{g_i} & , & ST_{g_{i+1}}^{min} & , & \overline{ST}_{g_{i+2}} \end{pmatrix}.$$

An illustration of the situation is given in Figure 5.11. With $CT_{g_{i+1}}^{min} < CT_{g_{i+1}}^{oldmin}$ we

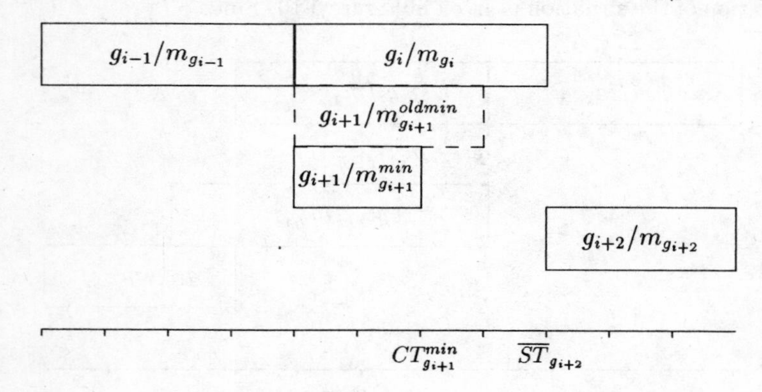

Figure 5.11: Illustration of Bounding Rule 5 – Case (b)

have $K_{rt}^{\rho}(\overline{\mathcal{PS}}_{i+1}) = K_{rt}^{\rho}(\mathcal{PS}_{i+1})$ for each period t, $t = CT_{g_{i+1}}^{oldmin} + 1, \ldots, \overline{T}$ and renewable resource r, $r \in R$. Since $\overline{ST}_{g_{i+2}} \geq CT_{g_{i+1}}^{oldmin} \geq ST_{g_{i+1}}^{oldmin}$ we have that $\overline{ST}_{g_{i+2}}$ is a feasible start time of activity g_{i+2} in mode $m_{g_{i+2}}$ with $(i+1)$-partial schedule $\mathcal{PS}_{i+1}$. Therefore, $ST_{g_{i+2}} \leq \overline{ST}_{g_{i+2}}$ which implies $K_{rt}^{\rho}(\overline{\mathcal{PS}}_{i+2}) \leq K_{rt}^{\rho}(\mathcal{PS}_{i+2})$ for each t, $t = \overline{ST}_{g_{i+2}} + 1, \ldots, \overline{T}$, and r, $r \in R$, and the proof is complete. □

5.6 Limitations of the Branch and Bound Procedure

In this section we will consider two extensions of the scheduling problem of type GRCPSP. First, the generalized temporal constraints (cf. Section 3.1) and second, the time-varying request for renewable resources (cf. Section 3.2). In both cases examples will illustrate that the foregoing as presented in Section 5.2 and Section 5.3 will only generate heuristic solutions, if the minimization of the makespan is considered.

The precedence tree related to the single-mode problem displayed in Table 5.8 with

j	d_j	$\mathcal{S}_j$	τ_{jh}	k_{j1}^{ρ}
1	0	$\{2,3\}$	$\tau_{12}^{min} = 0, \hat{\tau}_{12}^{max} = \infty$	0
			$\tau_{13}^{min} = 0, \hat{\tau}_{13}^{max} = \infty$	
2	3	$\{4\}$	$\tau_{24}^{min} = 0, \hat{\tau}_{24}^{max} = \infty$	3
3	1	$\{4\}$	$\tau_{34}^{min} = 0, \hat{\tau}_{34}^{max} = 3$	2
4	2	$\{\}$		4

Table 5.8: Counter Example Generalized Temporal Constraints

$\overline{T} = 6$, $|R| = 1$, $K_{1t}^{\rho} = 5$ for $t = 1, \ldots, 6$, has two branches to be evaluated; the branch 1-2-3-4 and the branch 1-3-2-4. Since the minimal time lags τ_{hj}^{min} are all equal to zero the critical path analysis presented in Section 1.3 does not change. Using the earliest precedence- and resource-feasible start times both branches would generate infeasible solutions. The optimal solution with a makespan of five periods is displayed in Figure 5.12.

Therefore: Before selecting another mode or descendant in Step 2 the next precedence and resource feasible start time $\overline{t} > ST_{g_i}$ has to be examined. The next mode (descendant) will only be selected if no more feasible start time can be found.

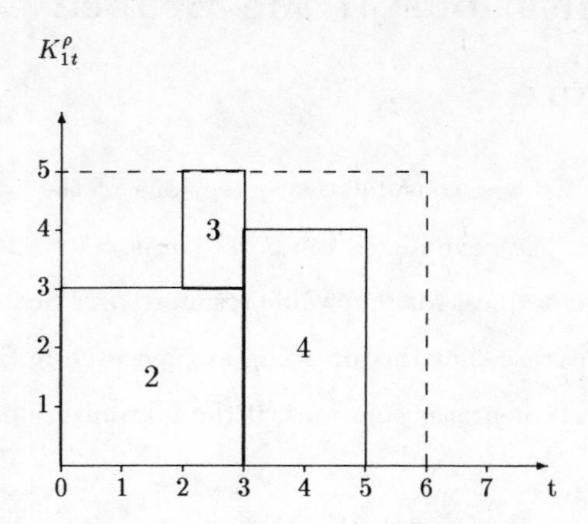

Figure 5.12: Solution for the Counter Example

However, if only job (project) specific release dates and deadlines have to be taken into account then the calculation of the time window can be simply adapted in order to obtain an optimal solution.

We now use the single-mode problem displayed in Table 5.9, where $\overline{T} = 5$, $|R| = 1$, $|D| = |N| = 0$ and k^ρ_{jt} denotes the usage of the renewable resource in the t'th period job j , $j = 1, \ldots, J$, $t = 1, \ldots, d_j$, is in process. Again, two branches of the related

j	d_j	$\mathcal{S}_j$	t	1	2	3	4	5
-	-		K^ρ_{1t}	1	2	4	4	4
1	0	$\{2,3\}$						
2	3	$\{4\}$	k^ρ_{2t}	1	2	4		
3	2	$\{4\}$	k^ρ_{3t}	1	2			
4	0	$\{\}$						

Table 5.9: Counter Example Time-Varying Request

precedence tree have to be evaluated, the branch 1-2-3-4 and the branch 1-3-2-4. We employ the scheduling strategy as presented in Table 5.4, Step 4 and yield the start time assignments $ST_1 = 0$, $ST_2 = 0$, $ST_3 = 3$, $ST_4 = 5$ and $\overline{ST}_1 = 0$, $\overline{ST}_2 = 2$, $\overline{ST}_3 = 0$, $\overline{ST}_4 = 5$. Clearly, both solutions are not optimal with respect to the makespan criterion. The optimal solution is given by $ST_1' = 0$, $ST_2' = 1$, $ST_3' = 1$, $ST_4' = 4$.

Again a "third" dimension for the determination of the start times has to be added, that is before switching to another mode or descendant the remaining feasible start times for the actual job/mode-combination have to be examined.

Chapter 6

Generation of Instances by ProGen

In this chapter we describe an algorithm for the generation of a general class of precedence- and resource-constrained scheduling problems. The generator has been developed by Kolisch, Sprecher and Drexl (cf. [68]). It has been coded in Turbo Pascal.

In order to show the effect of the bounding rules presented in Chapter 5 we will make use of the generator and the problem instances generated for the evaluation and validation of the project generator (ProGen).

Apart from minor changes in the introduction (Section 6.1) and the modified description of ProGen specific notation and symbols (Section 6.2) the problem generator is presented in [68]. For the computional studies which have been performed in order to validate the generator we refer to the original work.

6.1 Introduction

From the beginning of resource-constrained project scheduling research, rapid progress regarding models and methods has been documented in the literature (cf. [6], [7], [15], [25], [31], [32], [60], [88], [94], [113], [115], and [116]). But at the same time very little research concerned with the systematic generation of benchmark instances has been published. In [58] only a generator for random project scheduling problems is given. Unfortunately, it does not allow one to create instances subject to certain project characteristics. Hence, for experimental purposes, many researchers have generated their own test problems; sometimes utilizing a very restricted subset of project characteristics. Some of this work is rather well documented (cf. [25], [60], [69], [71], [95], [101], [112]), while most efforts are only briefly described (cf. [3], [13], [14], [18], [24], [26], [27], [41], [43], [70], [72], [74], [82], [83], [89], [98], [104], [114], [117], [119], and [124]). As a result, only a few commonly used benchmark instances are available. In 1984 Patterson compared four exact procedures for makespan minimization of the single-mode resource-constrained project scheduling problems (cf. [84]). These 110 problems have been (partially) used by [8], [9], [25], [28], [32] [66], [83], [86], [87], [98], [99] and [116] and therefore became a quasi standard. Nevertheless, there are some points of attack left:

- As a collection of problems from different sources, the problems are not generated by using a controlled design of specified parameters.
- Only the single-mode case and makespan minimization is taken into consideration.
- Recent advances (cf. [32]) in the development of exact single-mode procedures have demonstrated that the Patterson-set is solvable within an average CPU-time of less than a second on a personal computer. Since there are instances (with the same number of activities) which are much more difficult to solve, they can no longer be considered as a benchmark (cf. [68]).

The intention of this chapter is (cf. [59]): to present an instance generator for a broad class of project scheduling problems which utilizes several parameters. Some of them have been proposed in the former literature, others are entirely new.

The remainder of this chapter is organized as follows: In Section 6.2 we present ProGen specific notation and symbols. The employed parameters and their realization within ProGen are dealt with in Sections 6.3 and 6.4.

$p = 1, \ldots, P$	:	projects
ρ_p ($\overline{\delta}_p$)	:	release date (due date) of project p
FJ_p (LJ_p)	:	number of the first (last) job of project p
$j = 1$ (J)	:	unique source (sink) of the network
K_r^ρ	:	per period availability of renewable (doubly constrained) resource r
K_r^ν	:	total availability of nonrenewable (doubly constrained) resource r

Table 6.1: ProGen Specific Symbols and Definitions

6.2 ProGen Specific Notation and Symbols

We consider P projects, where each project has a specific release date ρ_p as well as a due date $\overline{\delta}_p$. The overall (super-)project consists of J partially ordered jobs, where $j = 1$ ($j = J$) is the unique dummy source (sink). For the sake of simplicity, project refers to the overall (super-) project as well. Furthermore, the jobs within the projects are consecutively labeled with FJ_p (LJ_p) being the first (last) job of project p. Thus project p consists of $LJ_p - FJ_p + 1$ jobs.

A resource r, $r \in R$, has a constant period capacity of K_r^ρ and each resource r, $r \in N$, has an overall capacity of K_r^ν units. Doubly constrained resources r, $r \in D$, are limited with respect to period capacity K_r^ρ and total capacity K_r^ν. Table 6.1 provides a summary of the notation and definitions.

6.3 Project Generation

I. Base Data Generation

In this subsection we briefly outline the generation of the projects base data. We use the functions round$(\cdot)$ and trunc$(\cdot)$, where the former rounds the argument to an integer and the latter truncates the decimal fraction of the argument. Furthermore, the random functions rand and $\overline{\text{rand}}$ are defined as follows:

$$\text{rand}[n_1, n_2] \;:\; \text{integer random number out of the interval } [n_1, n_2]$$
$$\overline{\text{rand}}[n_1, n_2] \;:\; \text{real random number out of the interval } [n_1, n_2]$$

The (pseudo) random numbers are constructed by transforming $[0, 1)$ uniformly distributed random numbers. The $[0, 1)$ uniformly distributed random numbers are calculated via the congruence-generator developed by Lehmer using the constants and implementation as given in [102].

The input and output for the generation of the base data is displayed in Tables 6.2 and 6.3, respectively. MPM_p denotes the MPM-duration of project p, $p = 1, \ldots, P$. It is calculated with respect to the release date by using the modes of shortest duration and the network, the construction of which is described in the next section. The due dates $\overline{\delta}_p$, $p = 1, \ldots, P$, are calculated as the truncation of the convex combination of the MPM-durations MPM_p and the horizon $\overline{T}$, where the due date factor δ_{fac}, $\delta_{fac} \in [0, 1]$, is used as the weight.

II. Network Generation

In Section 1.2 we stated that the structure of the project can be depicted as an acyclic activity-on-node network. Thus it is a quite natural approach to construct the network by using the following simple implication of the definition of a network:

P	:	number of projects
$J^{min}(J^{max})$	:	minimal (maximal) number of jobs per project
$M^{min}(M^{max})$	:	minimal (maximal) number of modes per job
$d^{min}(d^{max})$	:	minimal (maximal) duration per job
ρ^{max}	:	maximal release date
δ_{fac}	:	due date factor $\in [0,1]$

Table 6.2: Input Base Data Generation

Theorem 6.1 (cf. [79], p.33)

Let $N = (V, A)$ be a network with node set V and arc set A. Then, for every node $v \in V$ there is a directed path from the single source to v and a directed path from v to the single sink.

That is, every node except of the sink (source) has at least one successor (predecessor). Therefore the basic idea is as follows: First, determine one predecessor for each node, second, determine one successor for each node and then add further arcs. We consider the example in Figure 6.1 (cf. [49], p. 179), where the additional arc

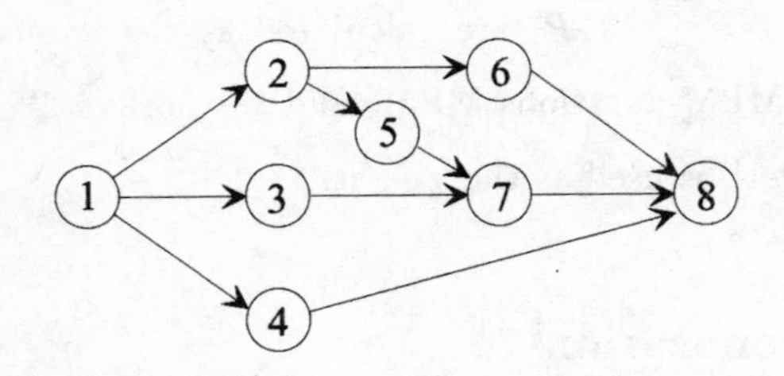

Figure 6.1: Example Network

$(2,7)$ would give no extra information about scheduling the activities and therefore should not be taken into consideration. We use the following definition:

J_p	$:=$	$\text{rand}[J^{min}, J^{max}]$, $p = 1, \ldots, P$
	$=$	number of jobs of project p
J	$:=$	$\sum_{p=1}^{P} J_p + 2$
	$=$	total number of jobs (including super-source and -sink). The jobs are numerically and consecutively labeled within the projects. That is, project p consists of the numerically labeled jobs j, $j = \sum_{q=1}^{p-1} J_q + 2, \ldots, \sum_{q=1}^{p} J_q + 1$.
M_j	$:=$	$\text{rand}[M^{min}, M^{max}]$, $j = 2, \ldots, J-1$ $(M_1 = M_J = 1)$
	$=$	number of modes of job j
d_{jm}	$:=$	$\text{rand}[d^{min}, d^{max}]$, $j = 2, \ldots, J-1$, $m = 1, \ldots, M_j$ and $d_{11} = d_{J1} = 0$. The modes are labeled with respect to non-decreasing durations.
ρ_p	$:=$	$\text{rand}[0, \rho^{max}]$
	$=$	release date of project p
$\overline{T}$	$:=$	$\max_{p=1}^{P} \rho_p + \sum_{j=1}^{J} \max_{m=1}^{M_j}\{d_{jm}\}$
	$=$	horizon
$\overline{\delta}_p$	$:=$	$\text{trunc}(\text{MPM}_p + \delta_{fac}(\overline{T} - \text{MPM}_p))$
	$=$	due date of project p

Table 6.3: Output Base Data Generation

Definition 6.1

Let $N = (V, A)$ be a network. An arc (i, j) is called redundant, if there are arcs $(i_0, i_1), \ldots, (i_{s-1}, i_s) \in A$ with $i_0 = i$, $i_s = j$ and $s \geq 2$.

That is, an arc (i, j) is redundant, if it is an element of the transitive closure $\overline{N}^+$ of $\overline{N} = (V, A \backslash \{(i, j)\})$. If within the construction process of the network an arc (i, j) is chosen for adding it to the current graph, four cases of redundancy might occur (cf. Figure 6.2), where $\overline{N} = (V, A)$ denotes the current graph with current sets of (immediate) successors $\overline{\mathcal{S}_j}(\mathcal{S}_j)$ and (immediate) predecessors $\overline{\mathcal{P}_j}(\mathcal{P}_j)$. For a given

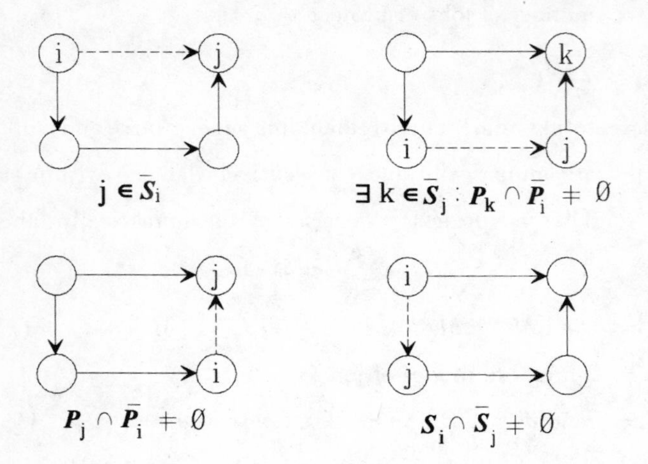

Figure 6.2: Cases of Redundancy

cardinality of the set of nodes the mimimal and maximal number of non-redundant arcs are given in the following theorem and illustrated in Figures 6.3 and 6.4.

Theorem 6.2

Let $N = (V, A)$ be a network with $|V| = n$.

(a) Since a network is connected, the minimal number of non-redundant arcs A^{min} is given by

$$A^{min} \quad = \quad n - 1.$$

(b) The maximal number of non-redundant arcs A^{max} in a network with $n \geq 6$ is given by

$$A^{max} \quad = \quad \begin{cases} n - 2 + \left(\frac{n-2}{2}\right)^2 & : \text{ if } n \text{ is even} \\ n - 2 + \left(\frac{n-1}{2}\right)\left(\frac{n-3}{2}\right) & : \text{ if } n \text{ is odd.} \end{cases}$$

For the characterization of the network we use the parameters given in Table 6.4. The complexity as the average number of (non-redundant) arcs per node is a measure

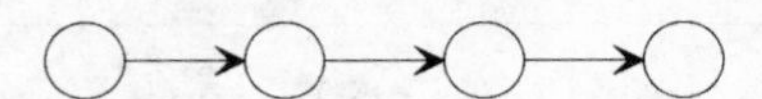

Figure 6.3: Minimal Number of Non-redundant Arcs

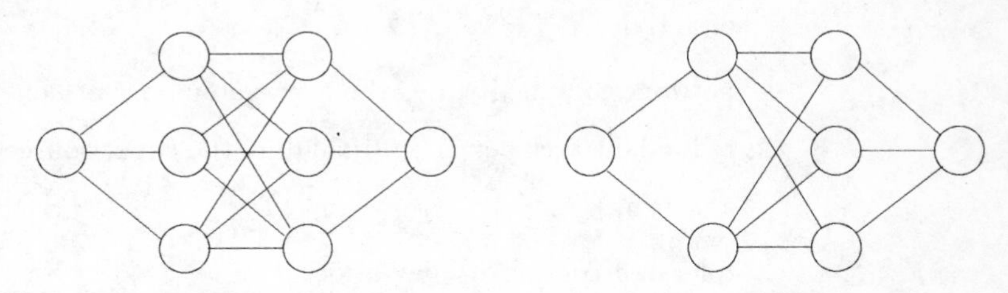

Figure 6.4: Maximal Number of Non-redundant Arcs

for the network logic, which has been introduced by Pascoe (cf. [82]) for activity-on-arc networks and adopted by Davis (cf. [26]) for the AON-representation. For the latter, complexity has to be understood such that for a fixed number of jobs, a higher complexity results in an increasing number of arcs and therefore in a greater interconnectedness of the network. It has already been shown by Alvarez-Valdes/Tamarit (cf. [3]), and will be confirmed in this study, that with increasing complexity problems become easier. This makes the term complexity somewhat confounding. Nevertheless, we stay with the term, because it has been used in a number of computational studies (cf. [43], [71], [83], [112], and [115]) and has become a wellknown project summary measure. Two disadvantages associated with this measure have to be mentionend - to wit:

(i) The number of arcs only does not give all informations about the number of possible schedules. Attempts to find more elaborate measures than complexity can be found in [51], [61] and [118]. But as pointed out by Elmaghraby and Herroelen (cf. [51]) "it seems evident to us that the structure of the network - in whichever way it is measured - will not be sufficient to reflect the difficulty encountered in the resolution of such problems".

$\mathcal{S}_1^{min}$ ($\mathcal{S}_1^{max}$)	:	minimal (maximal) number of start activities
$\mathcal{P}_J^{min}$ ($\mathcal{P}_J^{max}$)	:	minimal (maximal) number of finish activities
$\mathcal{S}_j^{max}$ ($\mathcal{P}_j^{max}$)	:	maximal number of successor (predecessor) activities of activity j, $j = 2, \ldots, J-2$
C	:	network complexity, i.e. the average number of non-redundant arcs per node (including the super-source and -sink)
ϵ_{NET}	:	tolerated complexity deviation

Table 6.4: Input Network Generation

(ii) The measure is not normalized to the interval [0,1]. A normalized measure for the network structure is the order strength which has been proposed by Mastor (cf. [75]) for the assembly line balancing problem and used by Cooper (cf. [23]) for the project scheduling problem. The order strength for the AON-representation is calculated by dividing the number of arcs associated with the transitive closure by the maximal number of arcs which is $n(n-1)/2$. Unfortunately the maximal number of arcs has two drawbacks: It includes redundant arcs and is far greater then the number of precedence relations occuring in scheduling problems. Although we can use the maximal number of non-redundant arcs for a normalizational purpose, this number still exceeds the number of precedence relations in a project. As a consequence, for projects, the order strength converges to zero with an increasing number of jobs.

We now describe the network construction for a single project (Figure 6.5), a multi-project network is maintained analogously. In Step 1 the number of start and finish activities are drawn randomly out of the interval $[\mathcal{S}_1^{min}, \mathcal{S}_1^{max}]$ and $[\mathcal{P}_J^{min}, \mathcal{P}_J^{max}]$, respectively. Arcs which connect the dummy source with the start activities and the finish-activities with the dummy sink are then added to the network. In Step 2,

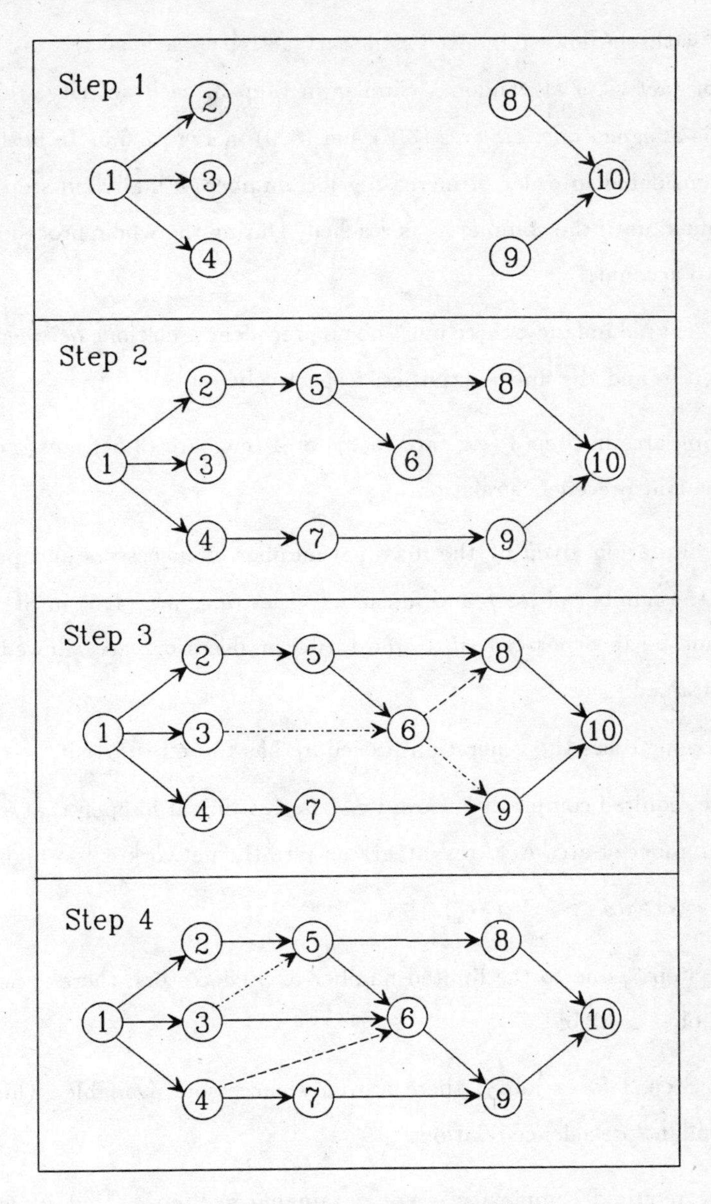

Figure 6.5: Network Generation

beginning with the lowest indexed non-start activity, each activity is assigned a predecessor (activity) at random. Similar in Step 3, each activity which has no successor is assigned one, cf. arcs $(3,6)$ and $(6,9)$ in Figure 6.5. In both steps the jobs are considered in order of increasing job number. Finally (in step 4) further arcs are added until the complexity is reached. During the whole procedure one has to take into account:

- To avoid redundancy, there must be no precedence relations between the start-activities and the finish-activities, respectively.
- Adding arcs in Step 3 (e.g. arc $(6,8)$) or 4 (e.g. arc (3,5)) must not produce redundant precedence relations.
- The limitation given by the maximal number of successors and predecessors and the number of start and finish activities (e.g. arc $(4,6)$ in Step 4, which cannot be incorporated, if at most two predecessors are allowed) must be maintained.

In the following cases the generation procedure has to be restarted:

- If the required complexity is low, i.e. $C \approx 1$, it might happen that after Step 3 the number of arcs ActArcs integrated into the network is too high, that is,

$$\text{ActArcs} \;>\; J \cdot C \cdot (1 + \epsilon_{NET}).$$

- If in Step 3, due to the limited number of predecessors, there is no successor of a job j available.
- If in Step 3 for a job j, there are only successors available, which lead to redundant precedence relations.
- If the required complexity is not obtainable in Step 4, that is, within a limited number of trials of randomly selecting a node and calculating possible successors, there are no further arcs addable to obtain

$$\text{ActArcs} \;\geq\; J \cdot C \cdot (1 - \epsilon_{NET}).$$

By an appropriate reduction of the set of choosable predecessors and successors in the steps previously described a numerically labeled network is realized.
Through adjustment of the input parameters, special network structures, e.g. general (Figure 6.6), serial structures (Figure 6.7) and network shapes as described in [70], [71] and [108] are obtainable.

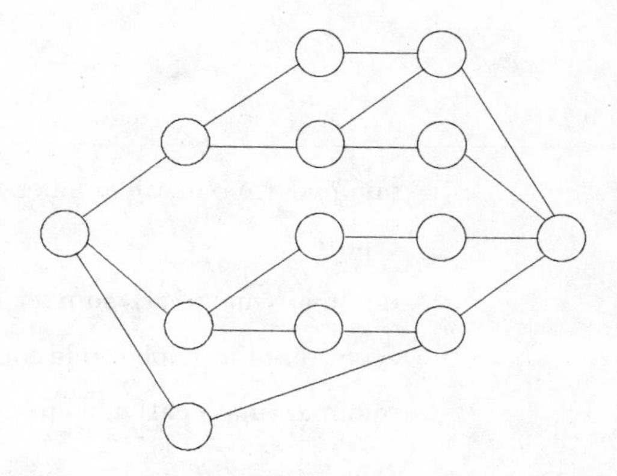

Figure 6.6: Multi-Project with General Structure

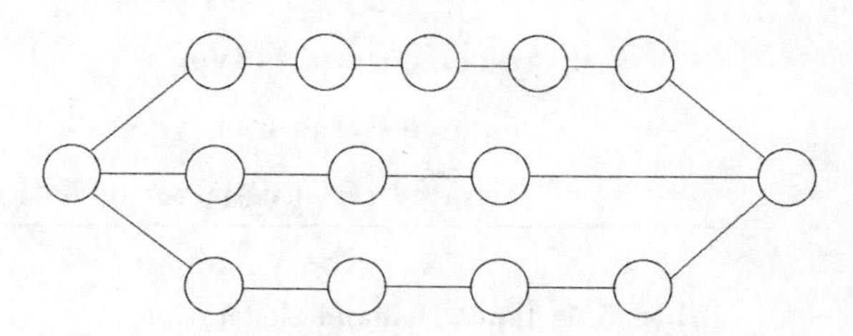

Figure 6.7: Multi-Project with Serial Structure

6.4 Resource Demand and Availability Generation

We consider a resource type τ, $\tau \in \{R, N, D\}$.

I. Resource Demand Generation

The parameters used in the demand generation are displayed in Table 6.5. The number of resources of type τ is determined by a randomly drawn integer within $[|\tau|^{min}, |\tau|^{max}]$, that is

$$|\tau| \ := \ \text{rand}[|\tau|^{min}, |\tau|^{max}].$$

$\lvert\tau\rvert^{min}(\lvert\tau\rvert^{max})$	:	minimal (maximal) number of resources of type τ
$Q_\tau^{min}(Q_\tau^{max})$	:	minimal (maximal) number of resources of type τ used by a job-mode combination $[j,m]$
$U_\tau^{min}(U_\tau^{max})$	:	minimal (maximal) demand for a resource of type τ
$P_\tau(F=1)(P_\tau(F=2))$	:	probability that demand for a resource of type τ is duration constant (monotonically decreasing with the duration)
RF_τ	:	resource factor of type τ
RS_τ	:	resource strength of type τ
ϵ_{RF}	:	tolerated resource factor deviation

Table 6.5: Input Demand Generation

The resource demand generation requires two decisions be made. First, we have to determine the resources used or consumed by the job-mode combinations $[j, m]$, $j = 1, \ldots, J$, $m = 1, \ldots, M_j$. Second, if a job-mode combination uses or consumes a resource, we have to calculate the number of units used or consumed. We refer to the first step as *request generation* and to the latter we refer to as *generation of demand level.*

Requested Resources

For characterizational purposes we use a generalization of the resource factor (RF) which has been introduced by Pascoe (cf. [82]) for the single-mode case and which has later on been utilized in studies by Cooper (cf. [23]) and Alvarez-Valdes/Tamarit (cf. [3]). For the single-mode case, RF is calculated as follows:

$$RF \quad := \quad \frac{1}{J}\frac{1}{|R|}\sum_{j=1}^{J}\sum_{r\in R}\begin{cases} 1 & , \quad \text{if } k_{jr} > 0 \\ 0 & , \quad \text{otherwise} \end{cases}$$

The resource factor reflects the average portion of resources requested per job. It is a measure of the density of the array k_{jr}. If we have $RF = 1$, then each job requests all resources. $RF = 0$ indicates that no job requests any resource, thus we obtain the unconstrained MPM-case. In order to use RF for the multi-mode case as well, we make a straightforward generalization to a type dependent resource factor RF_τ, $\tau \in \{R, N, D\}$:

$$RF_\tau \quad := \quad \frac{1}{J-2}\frac{1}{|\tau|}\sum_{j=2}^{J-1}\frac{1}{M_j}\sum_{m=1}^{M_j}\sum_{r\in\tau}\begin{cases} 1 & , \quad \text{if } k_{jmr} > 0 \\ 0 & , \quad \text{otherwise} \end{cases}$$

Again RF is normalized to the interval $[0,1]$ with the interpretation very close to that of the original RF. It reflects the average portion of resources out of one type, requested by each job-mode combination $[j,m]$ and it measures the density of the three dimensional array k_{jmr}. Of course, our RF equals the one proposed by Pascoe for the case $|N| = |D| = 0$ and $M_j = 1$, $j = 1,\ldots,J$.

For the generation of the resource request we use the following internal variables and data structures: First, we represent the information whether a job-mode combination $[j,m]$ requests resource r by a three-dimensional array $Rq[j,m,r]$ of binary digits. $Rq[j,m,r]$ is initialized with zeros and is set equal to one, if and only if $[j,m]$ requests resource r. The current resource factor (ARF) is then calculated as follows:

$$ARF_\tau \quad := \quad \frac{1}{J-2}\frac{1}{|\tau|}\sum_{j=2}^{J-2}\frac{1}{M_j}\sum_{m=1}^{M_j}\sum_{r\in\tau} Rq[j,m,r].$$

The current number of resources requested by $[j, m]$ is obtained by

$$\text{Q[j,m]} \;:=\; \sum_{r \in \tau} Rq[j, m, r].$$

Finally we get CT, the current set of choosable triplets,

$$CT \;:=\; \{[j, m, r]; Rq[j, m, r] = 0 \text{ and } Q[j, m] < Q_\tau^{max}\},$$

that is, the set of job-mode-resource combinations $[j, m, r]$, which are furthermore choosable ($Rq[j, m, r] = 0$) without $Q[j, m]$ exceeding Q_τ^{max}.

	Establishing the minimal number of resources requested by $[j, m]$			Establishing the resource factor		
2	1	…	M_2	1	…	M_2
	1 2 3		1 2 3	1 2 3		1 2 3
	0 0 1	…	1 0 0	1 0 1	…	1 0 1
	⋮		⋮	⋮		⋮
$J-1$	1	…	M_{J-1}	1	…	M_{J-1}
	1 2 3	…	1 2 3	1 2 3	…	1 2 3
	1 0 0	…	0 0 1	1 0 0	…	1 0 0
	Step 1			Step 2		
	$Q_\tau^{min} = 1, Q_\tau^{max} = 2$					

Table 6.6: Resource Factor Establishing

During the two steps to be performed the internal variables are continuously updated.

In Step 1 for each job-mode combination $[j, m]$, as long as the minimal number of requested resources Q_τ^{min} is not reached, additional resources are selected randomly. While, in Step 2, the current resource factor is less than the asserted one and in addition there are choosable triplets in CT, i.e. $CT \neq \emptyset$, the current resource factor is incremented by randomly drawing a triplet out of CT. In Table 6.6, where we have $|\tau| = 3$, the triplet (2,1,2) is not in the choosable set CT, because Q_τ^{max} is fixed to two.

If after Step 2 the current resource factor declines more then tolerated, i.e.

$$ARF_\tau \notin [RF_\tau \cdot (1 - \epsilon_{RF}), RF_\tau \cdot (1 + \epsilon_{RF})],$$

then a warning message is given.

Level of Demand

If we have $Rq[j, m, r] = 1$, then a positive demand of the job-mode combination $[j, m]$ for resource r has to be generated. The interrelation between the durations of the modes and the demand for resource r is reflected by two types of functions. One of which is duration independent ($F = 1$) and the other one is decreasing with the (increasing) duration ($F = 2$). That is, for the renewable and doubly constrained resources the per-period demand and for the nonrenewable resources the total demand is generated as the interrelation prescribes. For each resource $r \in \tau$ the interrelation is defined by

$$F_\tau(r) := \begin{cases} 1 & : \text{ if } \overline{\text{rand}}[0,1] < P_\tau(F = 1) \\ 2 & : \text{ otherwise} \end{cases}$$

given the type dependent probabilities $P_\tau(F = 1)$ and $P_\tau(F = 2)$. If $F_\tau(r) = 1$, then for each job the demand U' is randomly drawn out of the integer interval $[U_\tau^{min}, U_\tau^{max}]$ and is then assigned to all modes, which request this resource. In the case of $F_\tau(r) = 2$, for each job j two levels are drawn randomly out of the parameter specified interval:

$$U^1 := \text{rand}[U_\tau^{min}, U_\tau^{max}] \quad , \quad U^2 := \text{rand}[U_\tau^{min}, U_\tau^{max}].$$

Then U^{low} and U^{high} are calculated as follows

$$U^{low} := \min\{U^1, U^2\} \quad , \quad U^{high} := \max\{U^1, U^2\}$$

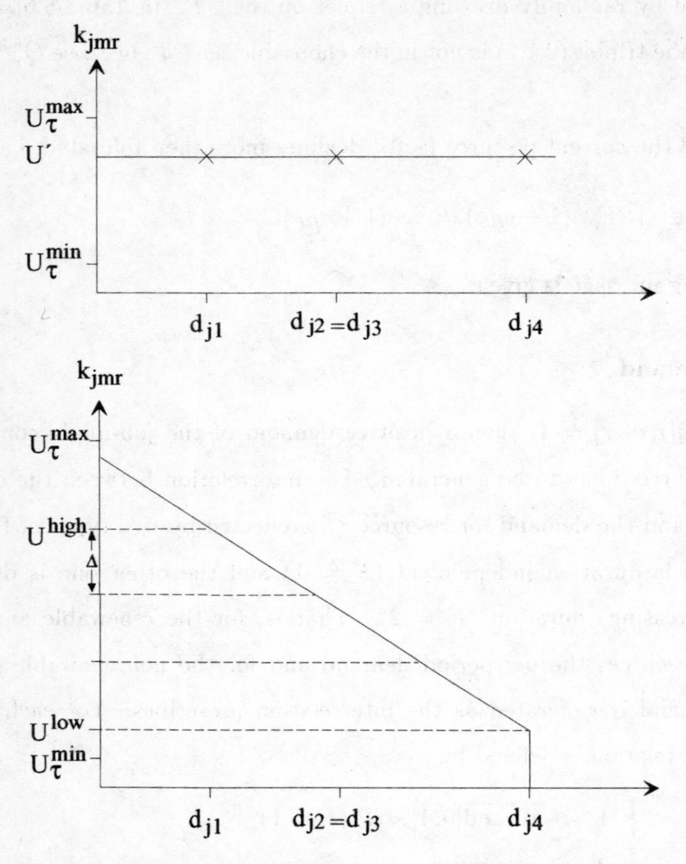

Figure 6.8: Interrelation between Demand and Duration

Let $\overline{M_j}$ be the number of modes of job j with different durations requesting resource r. We calculate

$$\Delta \;:=\; \frac{U^{high} - U^{low}}{\overline{M_j}}$$

and yield $\overline{M_j}$ intervals I_k as follows:

$$I_k := [round(U^{high} - \Delta k), round(U^{high} - \Delta(k-1))] \qquad k = 1, \ldots, \overline{M_j}.$$

Since the modes are labeled with respect to nondecreasing durations, we can now draw the demand randomly out of the intervals corresponding to the durations.

Figure 6.8 illustrates the generation of the level of demand.

Remark 6.1

If for $m, \overline{m} \in \{1, \ldots, M_j\}$, $m \neq \overline{m}$, it is $d_{jm} = d_{j\overline{m}}$ and $Rq[j,m,r] = 1 = Rq[j,\overline{m},r]$, then the demand is generated randomly out of the same interval.

Due to the construction inefficiency, which is defined in the following, might occur:

Definition 6.2

A job j has inefficient modes, if there are modes m and $\overline{m}$ with $d_{jm} \leq d_{j\overline{m}}$ and $k^{\rho}_{jmr} \leq k^{\rho}_{j\overline{m}r}$ for all $r \in R \cup D$ and $k^{\nu}_{jmr} \leq k^{\nu}_{j\overline{m}r}$ for all $r \in N \cup D$.

If inefficient modes occur for job j, we calculate the number of resources requested by job j

$$Q_j \;:=\; \sum_{m=1}^{M_j} \sum_{r \in \tau} Rq[j,m,r]$$

and the request and demand generation is restarted with the additional constraint

$$Q_j \;=\; \sum_{m=1}^{M_j} Q[j,m].$$

If efficiency is not obtainable within MaxTrials, the generation is interrupted and the parameters have to be adjusted.

II. Resource Availability Generation

In order to express the relationship between the resource demand of the jobs and the resource availability Cooper (cf. [23]) introduced the resource strength (RS), which is calculated as follows:

$$RS_r := \frac{K_r}{\frac{1}{J}\sum_{j=1}^{J} k_{jr}}$$

Later, the RS has been utilized by Alvarez-Valdes/Tamarit (cf. [3]). There are three main drawbacks of the proposed measure. We will point them out and propose a new RS to overcome these disadvantages:

- First, the RS is not standardized in the intverval $[0, 1]$.

- Second, a rather small RS does not guarantee a feasible solution. E.g. for three jobs with $k_{jr}^{\delta} = 1$, 1 and 10, respectively, one has to adjust the resource strength to $RS_r \geq 2.5$ in order to achieve a feasible solution.

- Third and most important, consider the myopic fashion in which the scarcity of resources is calculated. This shall be depicted with the following simple example: We consider two projects, with exactly the same data except the network. Project 1 has a parallel structure, where each job is an immediate successor of the dummy source and an immediate predecessors of the dummy sink, whereas project 2 has a serial structure, where each job has exactly one predecessor and one successor. Let us further assume that the resource availability is large enough to assure feasibilty of both problems. Then the RS for both projects will be exactly the same, but obviously the serially structured project, being the MPM-case, will be quite easy to solve, whereas the parallel structured project is, dependent on the amount of resource availability, rather difficult.

In order to overcome these disadvantages, we have created the following methodology for a measure of resource scarceness which is applicable to all types of resources. We determine a minimal demand K_r^{min} as well as a maximal demand K_r^{max} and let the resource availability be a convex combination of the two with RS_r as scaling parameter : $K_r := K_r^{min} + RS_r(K_r^{max} - K_r^{min})$. Thus with respect to one resource we will get the smallest feasible resource availabilty for $RS_r = 0$. For $RS_r = 1$ the amount of resources is approximately large enough to achieve the MPM-case.
For the nonrenewable resources r, $r \in N \cup D$, the minimal and maximal availabilities to complete the project can be calculated as follows:

$$K_r^{min} := \sum_{j=2}^{J-1} \min_{m=1}^{M_j} \{k_{jmr}^{\nu}\}$$

$$K_r^{max} := \sum_{j=2}^{J-1} \max_{m=1}^{M_j}\{k_{jmr}^{\nu}\}$$

For a given type dependent resource strength $RS_\tau \in [0,1]$ the availability is

$$K_r^{\nu} := K_r^{min} + \text{round}(RS_\tau\ (K_r^{max} - K_r^{min})).$$

If the considered resource is renewable the minimal demand is

$$K_r^{min} := \max_{j=2}^{J-1} \left\{ \min_{m=1}^{M_j}\{k_{jmr}^{\rho}\} \right\}.$$

The maximal demand is calculated as the peak demand of the precedence preserving the earliest start schedule. Each job is thereby performed in the lowest indexed mode employing maximal per-period demand with respect to the resource under consideration. That is, we determine the maximal per-period demand of job j with respect to resource r

$$k_{jr}^* := \max_{m=1}^{M_j}\{k_{jmr}^{\rho}\}$$

and the corressponding mode with shortest duration:

$$m_{jr}^* := \min_{m=1}^{M_j}\{m | k_{jmr}^{\rho} = k_{jr}^*\}$$

Given the precedence relations and release dates of the project, we can now calculate the earliest start schedule with the modes determined. We obtain the resource dependent start time ST_j^r and completion time CT_j^r of job j, $j = 2, \ldots, J-1$. We then calculate the peak period demand

$$K_r^{max} := \max_{t=1}^{\overline{T}} \left\{ \sum_{\substack{j=2 \\ ST_j^r+1 \le t \le CT_j^r}}^{J-1} k_{jm_{jr}^* r} \right\}$$

and the available amount using the type dependent resource strength RS_τ

$$K_r^{\rho} := K_r^{min} + \text{round}(RS_\tau(K_r^{max} - K_r^{min})). \tag{6.1}$$

By constuction we can state the following:

Remark 6.2

(a) If $|\tau| = 1$ and $RS_\tau = 0$, then the lowest resource feasible level with respect to τ will be generated.

(b) IF $RS_\tau << 1$ and $M_j > 1$, then feasibility of the problem can not be assured, because of mode coupling via resource constraints.

Chapter 7

Computational Results

In this chapter we present the results of our computational studies. One of the main results will be that, although the efficieny of the algorithm has been substantially increased by the proposed bounding rules, the multi-mode resource-constrained project scheduling problem is less tractable than reported in the literature. Patterson et al. (cf. [89]) have generated 91 instances. The number of jobs ranged from 10 to 500, where 75 instances have up to 30 jobs. The instances have been characterized by the mean number of modes, mean activity duration, minimum/maximum activity duration, standard deviation of the activity durations, critical path length (based on minimum activity durations), average fraction of resources used by an activity mode and network density. The procedure has been coded in Fortran and implemented on an IBM 4381 mainframe, which is, as has been stated, approximately seven times faster than a 386-based, 20 MHz PC with numeric coprocessor. For an imposed time limit of 1 (10) minutes 30 (33) of the problems with up to 50 activities have been solved to optimality. The preponderance of these problems ranged between ten and thirty jobs.

Throughout the chapter the objective under consideration is the minimization of the makespan. The evaluations of the experiments have been performed by the use of SPSS. The solution times and average solution times are given in seconds.

Since the computations have been performed on different machines we firstly present the solution times of a series of problem instances (cf. Table 7.1)

Producer	IBM	Noname	IBM	IBM	IBM	IBM
Type	PS2/55sx	comp.	PS2/70	PS2/95	RS6000	RS6000
System	DOS	DOS	OS/2	SCO-UNIX	AIX	AIX
Processor	80386sx	80386dx	80486dx	XP486	320	550
Clockpulse	16Mhz	40MHz	25Mhz	33Mhz	25MHz	41.6MHz
Language	MS-C	MS-C	MS-C	CC	CC	CC
1	275.07	70.58	99.69	20.06	25.87	13.51
2	56.13	14.50	21.09	4.42	5.53	2.84
3	662.18	169.67	241.75	49.76	62.73	32.84
4	87.60	22.58	33.16	6.96	9.03	4.51
5	1227.64	316.26	453.91	94.02	125.39	64.26
6	3185.08	817.79	1169.25	244.79	309.81	158.83
7	915.28	236.78	344.31	71.58	90.49	45.12
8	105.73	27.35	40.22	8.44	11.02	5.53
9	222.56	57.40	83.47	17.70	23.39	11.66
10	147.86	38.55	56.06	11.78	15.01	7.54
11	419.74	108.91	159.44	34.38	46.52	22.66
12	299.01	77.28	111.94	23.81	30.91	15.38
13	1193.09	305.99	450.31	90.65	125.95	63.05
14	572.10	147.80	215.63	45.90	58.54	29.78
15	465.33	119.46	173.68	35.94	47.23	24.08
16	346.80	90.57	132.94	27.87	37.18	18.66
17	250.24	64.92	96.29	20.09	26.50	14.12
18	77.39	20.16	29.66	6.38	8.26	4.32
19	325.71	84.81	125.85	26.38	32.60	17.99
20	568.42	145.55	211.25	43.77	56.37	31.15
$\sum$	11402.96	2936.91	4249.87	884.68	1148.33	587.83
Factor	19.40	5.00	7.23	1.50	1.95	1.00

Table 7.1: Comparison of Different Machines

solved with the algorithm of Table 5.4.

The sample has been constructed by ProGen. The instances we have used consist of twelve activities (including the dummy source and the dummy sink), the non-dummy activities can be performed in three modes, where two renewable and two nonrenewable resources are used and consumed, respectively. In order to find the optimal solution the procedure described in Table 5.4 has been employed. The basic version of the algorithm has been accelerated by the use of pointer arithmetic which is actually approximately two times faster than the counterpart employing array arithmetic.

Twenty instances have been generated and solved on the machines described in Table 7.1. Beside the solution times (in seconds) for each of the twenty instances and machine, the sum of the solution times for each machine is displayed in the table. The comparison factor is given in the last line. Nevertheless, one has to keep in mind that the algorithm proposed only makes use of integer arithmetic.

7.1 Exact Methods

In order to compare the effect of the priority rules (cf. Section 5.4) for the relabeling of the eligible set we have made use of the instances generated for the evaluation and validation of ProGen (cf. [68]). The constant and variable parameter settings are displayed in Table 7.2 and Table 7.3, respectively. We consider only single-project problems, where, recall, $J^{min} = J^{max} = 10$ produces a project which consists of ten non dummy activities.

Moreover, throughout the chapter, we have used only functions reflecting a decreasing level of usage (consumption) with increasing duration, i.e. $P_R(F = 2) = 1$ ($P_N(F = 2) = 1$). Furthermore, we have employed the tolerances ϵ_{NET} and ϵ_{AVL} of 0.05. The maximal number of trials has been fixed to 200.

	J	M_j	d_j	$\|R\|$	U_R	Q_R	$\|N\|$	U_N	Q_N	$\mathcal{S}_1$	$\mathcal{S}_j$	$\mathcal{P}_J$	$\mathcal{P}_j$
min	10	3	1	2	1	1	2	1	1	3	1	3	1
max	10	3	10	2	10	2	2	10	2	3	3	3	3

Table 7.2: Constant Parameter Levels for the Multi-Mode Instances under Full Factorial Design

Parameter	Levels			
RF_R	0.5	1.0		
RS_R	0.2	0.5	0.7	1.0
RF_N	0.5	1.0		
RS_N	0.2	0.5	0.7	1.0

Table 7.3: Variable Parameter Levels for Multi-Mode Instances under Full Factorial Design

Ten instances have been generated for each combination of the parameters. As outlined in [68] feasibility of all the problems cannot be guaranteed. Only 536 of the 640 problems have a feasible solution.

	Priority-Rule							
	P1	P2	P3	P4	P5	P6	P7	P8
μ_{CPU}	74.10	57.86	77.49	76.26	80.25	78.86	77.62	83.35
σ_{CPU}	214.42	163.42	222.28	221.06	228.81	224.58	221.88	242.16

Table 7.4: Effects of Priority Rules 1 to 8 (RS6000/320)

Table 7.4 shows the average computation times μ_{CPU} and the standard deviations σ_{CPU} for the algorithm of Table 5.4 employing the priority rules (cf. Table 5.6) in order to relable the eligible set. The related variants are denoted as P1,...,P8. The frequency distributions of the solution times are given in Table 7.5.

Rule	[0,0.1]	(0.1,1]	(1,5]	(5,10]	(10,25]	(25,50]	(50,100]	(100,250]	>250
P1	157	68	95	37	39	32	25	29	54
P2	155	71	106	30	47	27	21	38	41
P3	144	73	99	42	36	30	28	29	55
P4	149	68	97	43	37	37	22	27	56
P5	140	72	95	43	43	34	23	29	57
P6	146	73	88	49	38	30	26	27	59
P7	143	77	91	46	40	29	25	28	57
P8	156	66	87	40	42	29	32	27	57

Table 7.5: Frequency Distribution of Solution Times for Priority Rules 1 to 8 (RS6000/320)

It can be observed that on basis of average CPU-time the job number rule (rule 1) outperforms all the other priority rules but the random relabeling of the eligible set (rule 2). This is partly reasoned by the additional effort required for sorting the activities of the eligible set. One has to take into account that the effort of sorting is linear (with respect to the number of activities in the eligible set) only if randomly relabeling is used. In contrast, if the job number (rule 1), the minimum latest finish time (rule 5) or the minimum possible start time rule (rule 8) is used, then the successors of activity Y_{iN_i} have to be assigned a label which is greater than or equal to N_i. More precisely, the priority values of the new descendants have to be compared with the priority values of the activities out of Y_i having a label which is higher than N_i. If the maximum duration (rule 3), maximum average duration (rule 4), the minimum latest finish time reduced by maximum duration (rule 6) or the minimum latest finish time reduced by average duration (rule 7) rule is used then all the eligible acitvities have to be rearranged.

Further computational studies have turned out that omitting the job number rule saves approximately five percent of the CPU-time, if the new descendants are only appended to the set $Y_i \backslash \{g_i\}$.

For the following reasons we restrict ourselves for the rest of the investigation to the job number rule: First, additionally performed experiments have shown that the effect of randomly relabeling on the computational time can be more than twenty percent worse than the results obtained by employing the job number rule. Second, there are only marginal differences between the frequency distributions of CPU-time of the randomly relabeling and the job number rule. This supports the idea that there is no systematic effect by the use of the former priority rule. Third, the second bounding rule is not immediately applicable.

In the second experiment we have made use of the examples described above in order to find out the effect of the bounding rules presented in Section 5.5. All the problems have been solved by the variants listed in Table 7.6 where (+) indicates that the concerned rule is employed and (-) that it is not. Recall, bounding rule 1 induces backtracking, if an activity is not schedulable in any mode. Bounding rule 2 skips a job/mode combination $[g_{i+1}, m_{g_{i+1}}]$, if interchanging the scheduling sequence with the job/mode combination $[g_i, m_{g_i}]$ results in the same start times. Bounding rule 3 is the left-shift rule. Bounding rule 4 recognizes that a partial schedule is not completable with respect to nonrenewable resources. If only renewable resources are taken into consideration, then bounding rule 5 excludes mode changes, if they are without effect for the objective function under consideration.

Variant	Algorithm	Bounding Rule				
		1	2	3	4	5
V1	Algorithm of Table 5.4	–	–	–	–	–
V2	Algorithm of Table 5.4	+	–	–	–	–
V3	Algorithm of Table 5.4	–	+	–	–	–
V4	Algorithm of Table 5.4	–	–	+	–	–
V5	Algorithm of Table 5.4	–	–	–	+	–
V6	Algorithm of Table 5.4	+	+	+	+	–
V7	Algorithm of Table 5.4	+	+	+	–	–
V8	Algorithm of Table 5.4	+	+	+	–	+

Table 7.6: Variants of the Original Algorithm

	Variant					
	V1	V2	V3	V4	V5	V6
μ_{CPU}	74.10	63.53	22.27	22.49	6.57	0.71
σ_{CPU}	214.42	195.64	66.12	68.38	13.89	2.32
max_{CPU}	2908.27	2854.45	658.56	889.59	121.27	37.24
Factor	1.00	1.17	3.33	3.29	11.28	104.37

Table 7.7: Effects of Bounding Rules 1 to 4 (RS6000/320)

Table 7.7 and Table 7.8 depict the average solution times μ_{CPU}, the standard devia-

Variant	[0,0.1]	(0.1,1]	(1,5]	(5,10]	(10,25]	(25,50]	(50,100]	(100,250]	>250
V1	157	68	95	37	39	32	25	29	54
V2	171	76	93	34	30	34	18	33	47
V3	174	103	91	37	41	22	27	37	4
V4	153	92	107	37	40	33	42	22	10
V5	161	82	141	58	56	25	12	1	–
V6	248	217	55	11	4	1	–	–	–

Table 7.8: Frequency Distribution of Solution Times for Variants V1 - V6 (RS6000/320)

tions σ_{CPU}, the maximum computation times max_{CPU} and the frequency distributions of the variants V1-V6, respectively. The strongest effect has been obtained by bounding rule 4 which is due to the fact that the rule has been efficiently realized via preprocessing. Note, since this rule checks completability it is applicable to any objective function under consideration. In comparison to a previous version the effect of bounding rule 1 is slightly reduced. In the former realization of the algorithm we have used the time window $[\, t^*, LS_{g_i}]$ for the assignment of a start time to a job g_i in mode m_{g_i}, which clearly is different from using the interval $[t^*, LF_{g_i} - d_{g_i m_{g_i}}]$. In the former version it might happen that a time window violation is firstly identified on a later stage than on the current. The rules 2 and 3 have a similar effect, since both rules are applicable if sufficient units of renewable resources are available allowing more parallelism of the activities to be performed. From Table 7.8 it can be observed that if all the bounding rules are employed together (V6) then approximately 97% of the test problems are solvable within a CPU-time of up to five seconds. Whereas the original algorithm has a maximal CPU-time of 2908.27 seconds the accelerated one requires at most 37.24 seconds. Moreover, the accelerated version is on average more than one hundred times faster than the original one. It has to be mentioned that the algorithm has been slowed down by the bounding rules for only a few instances out of the lowest class ranging from 0.0 to 0.1 seconds.

Table 7.9 displays the average computation time for the different levels of RF_R, RS_R, RF_N and RS_N. For the calculation of the average CPU-times we have fixed the parameter under consideration and left the others free. For each level the number of feasible problems is given in the third column. The comparison factor of the variant V1 and V6 is depicted in the last column. The following observations can be made: The effect of the bounding rules is the stronger the lower the resoure factor and the higher the resource strength of the renewable resources are. For the nonrenewable resource the reverse is true. The effect of the rules decreases with the resource strength and increases with the resource factor.

The average computation times of variant V6 are the highest in the classes with

		Feasible	Variant						Factor
			V1	V2	V3	V4	V5	V6	
RF_R	0.5	259	62.23	54.50	17.04	13.04	5.32	0.26	239.35
	1.0	277	85.20	71.10	27.16	33.27	7.74	1.13	75.00
RS_R	0.2	119	65.63	52.92	34.07	44.59	11.46	2.37	27.69
	0.5	139	50.10	41.45	14.55	17.72	4.84	0.35	142.86
	0.7	138	89.10	78.75	21.01	19.82	5.62	0.20	445.50
	1.0	140	90.36	79.65	21.17	14.94	5.08	0.15	602.40
RF_N	0.5	232	3.83	2.40	1.71	2.62	3.68	0.64	5.98
	1.0	304	127.73	110.19	37.96	39.43	8.78	0.76	168.06
RS_N	0.2	76	441.32	383.61	130.62	128.23	12.89	0.79	621.58
	0.5	153	31.37	25.61	9.34	12.76	9.93	0.96	32.68
	0.7	156	6.48	4.92	2.39	3.78	4.22	0.57	11.37
	1.0	151	2.43	1.41	1.39	2.03	2.41	0.60	4.05

Table 7.9: Full Factorial Design – J=10 (RS6000/320)

a high resource factor RF_R and a low resource strength RS_R. This is reasoned by the lack of bounding rules especially designed for bounding the enumeration in this classes.

This gap might be closed by introducing for each renewable resource a nonrenewable one with availability and consumption in accordance with Remark 5.8 and applying bounding rule 4.

Clearly, it might happen that there is no effect of the rule, but we can enhance it by the following strategy: Let $\mathcal{PS}_i$ denote an i-partial schedule. We consider a job/mode combination $[g_{i+1}, m_{g_{i+1}}]$ with start time $ST_{g_{i+1}}$ on stage $i+1$. Let $\hat{T}$, $ST_{g_{i+1}} \leq \hat{T} \leq \overline{T}$, be a reduced horizon and let $\overline{\mathcal{AS}}$ denote the set of currently unscheduled activities. We determine the set $CT_{\hat{T}}$ of activities which have to be completed up to period $\hat{T}$, that is,

$$CT_{\hat{T}} \quad := \quad \{j \in \overline{\mathcal{AS}}; LF_j \leq \hat{T}\}$$

and define

$$\hat{\hat{T}} \;:=\; \max\{LF_j; j \in CT_{\hat{T}}\}.$$

Now we can calculate the total number of units $Ktot_r^\rho(\mathcal{PS}_i, ST_{g_{i+1}}, \hat{\hat{T}})$ of resource r, $r \in R$, available to complete the activities out of $CT_{\hat{T}}$, that is,

$$Ktot_r^\rho(\mathcal{PS}_i, ST_{g_{i+1}} \hat{\hat{T}}) \;:=\; \sum_{t=ST_{g_{i+1}}+1}^{\hat{T}} K_{rt}^\rho(\mathcal{PS}_i).$$

If for any renewable resource r, $r \in R$, the remaining capacity $Ktot_r^\rho(\mathcal{PS}_i, ST_{g_{i+1}}, \hat{\hat{T}})$ is less than the minimally required number of units

$$kmin_r^\rho(CT_{\hat{T}}) \;:=\; \sum_{j \in CT_{\hat{T}}} \min_{m=1}^{M_j}\{k_{jmr}^\rho\, d_{jm}\},$$

then the schedule $\mathcal{PS}_{i+1}$ is not completable and the next job/mode combination has to be tested on stage $(i+1)$.

The problem is the (iterative) determination of $\hat{T}$. $\hat{T}$ should be large enough to guarantee that $CT_{\hat{T}}$ is not empty and not too large to fathom a node in the Branch and Bound tree.

Furthermore, one can relax the renewable resource constraints by substituting all the renewable resources by a single renewable resource with per period availabilities of

$$K_t^\rho := \sum_{r \in R} K_{rt}^\rho, \qquad t = 1, \ldots, \overline{T} \tag{7.1}$$

and usages of

$$k_{jm}^\rho := \sum_{r \in R} k_{jmr}^\rho, \qquad j = 1, \ldots, J, m = 1, \ldots, M_j. \tag{7.2}$$

The optimal solution of the newly derived problem provides a lower bound on the objective function value of the original problem. Therefore, the algorithm can be stopped if the bound is reached or if the quality of the current solution fulfills predefined requirements.

In addition we have performed another full factorial design with $J = 12$ instead $J = 10$. Again, ten instances have been generated for each parameter configuration (cf. Table 7.2 and 7.3). The average CPU-times and the frequency distributions of the solution times are displayed in Table 7.10 and Table 7.11, respectively.

		Feasible	Variant V1	V6	Factor
RF_R	0.50	265	976.84	0.79	1236.50
	1.00	282	1584.81	6.09	260.23
RS_R	0.25	134	1228.21	11.74	104.62
	0.50	141	1306.31	1.55	842.78
	0.75	140	1466.62	0.58	2528.65
	1.00	132	1147.61	0.38	3020.02
RF_N	0.50	241	47.42	2.64	17.96
	1.00	306	2268.48	4.21	538.83
RS_N	0.25	72	8835.22	5.29	1670.17
	0.50	158	374.48	4.92	76.11
	0.75	158	41.62	2.74	15.19
	1.00	159	23.28	2.06	11.30
μ_{CPU}		547	1289.91	3.52	366.45
σ_{CPU}			4552.62	195.64	
max_{CPU}			51786.22	151.25	

Table 7.10: Full Factorial Design – J=12 (RS6000/320)

Variant	[0,0.1]	(0.1,1]	(1,5]	(5,10]	(10,25]	(25,50]	(50,100]	(100,250]	>250
V1	143	42	56	37	31	30	35	40	133
V6	196	166	118	25	25	10	5	2	–

Table 7.11: Frequency Distribution of Solution Times for Variant V1 and V6 (RS6000/320)

Only 547 of the 640 problems have a feasible solution. The number of feasible problems for the different levels of RF_R, RS_R, RF_N and RS_N is displayed in the third column of Table 7.10. The fourth and fifth column show the average CPU-time for optimally solving the problems of the concerned class by the variants V1 and V6, respectively. The comparison factor of the variant V1 and V6 is given in the last column. The maximum CPU-time for variant V1 and V6 has been 51786.22 and 151.25 seconds, respectively. The relationship between the resource factor and resource strength of the renewable and nonrenewable resources and the comparison factor of the variants V1 and V6 is the same as for $J = 10$. The comparison factor on the (overall) average CPU-time of the variants V1 and V6 is 366.45 instead of 104.37. The third experiment we have performed has been a ceteris paribus design in order to find out the effect of an increasing number of jobs, number of modes, number of start activities, complexity and number of renewable and nonrenewable resources on the solution times. The constant parameter setting is displayed in Table 7.12. Moreover, we have choosen RF_R, RS_R, RF_N and RS_N as 0.5.

	d_j	U_R	Q_R	U_N	Q_N	$\mathcal{S}_j$	$\mathcal{P}_J$	$\mathcal{P}_j$
min	1	1	1	1	1	1	3	1
max	10	10	$\|R\|$	10	$\|N\|$	3	3	3

Table 7.12: Constant Parameter Levels for the Ceteris Paribus Design

The variable parameter setting is depicted in Table 7.13, where the underbar denotes the standard setting.

For each configuration we have generated ten problems, all of which have a feasible solution. These instances have been solved to optimality by the variants V1 and V6. The average computation times and the comparison factors are given in Table 7.14. One can notice that due to the NP-hardness of the problem (cf. [64]) the average CPU-time substantially increases with the number of jobs. Similar for the number of modes. With an increasing number of start activities the average time to solve the problems increases due to the more parallelism of the structure of the network. If the number of precedence relations increases, then the problems become easier

J	10	12	14	16	
M_j	1	2	3	4	5
$\|\mathcal{S}_I\|$	1	2	3	4	
C	1.5	1.8	2.1		
$\|R\|$	2	3	4	5	
$\|N\|$	2	3	4	5	

Table 7.13: Variable Parameter Levels for the Ceteris Paribus Design

Parameter		V1	V6	Factor	Parameter		V1	V6	Factor
J	10	12.71	0.69	18.42	M_j	1	0.53	0.05	10.60
	12	80.15	1.64	48.48		2	15.43	0.55	28.05
	14	658.94	6.55	100.60		3	35.53	0.73	48.67
	16	15984.50	73.54	217.36		4	77.17	2.02	38.20
						5	383.53	10.94	35.05
$\|\mathcal{S}_I\|$	1	12.10	0.57	21.23	C	1.5	215.20	3.34	64.43
	2	21.34	1.27	16.80		1.8	135.06	1.56	86.58
	3	55.56	1.86	29.87		2.1	15.48	0.47	32.93
	4	205.48	2.51	81.86					
$\|R\|$	2	175.08	2.94	59.55	$\|N\|$	2	33.94	1.19	28.52
	3	236.70	4.29	55.17		3	115.14	2.46	46.80
	4	428.19	5.26	81.40		4	322.48	3.91	82.48
	5	506.74	11.56	43.84		5	415.41	5.34	77.79

Table 7.14: Average Computation Times for the Ceteris Paribus Design (80486)

to solve because the number of sequences to be examined is reduced. Moreover, the more resources have to be taken into account the more time has to be spent on solving the problem.

For the last experiment we have constructed scheduling problems where only renewable resources have be taken into consideration. The constant parameter setting is displayed in Table 7.15. The variable parameter levels and the average CPU-times within the classes are given in Table 7.16 and Table 7.17. Beside the resource factor and the resource strength we have varied the number of activities. The projects consist of 10, 12, 14 and 16 activities. Ten problems have been generated for each parameter combination. The problems have been solved with the variants V7 and V8. The number of feasible problems is given in the fourth and nineth column of the tables. The columns five, six, ten and eleven show the average computation times within the classes obtained by the variants V7 and V8. The comparison factors are depicted in row seven and twelve of the tables.

	M_j	d_j	$\|R\|$	U_R	Q_R	$\|N\|$	U_N	Q_N	$\mathcal{S}_1$	$\mathcal{S}_j$	$\mathcal{P}_J$	$\mathcal{P}_j$
min	3	1	2	1	1	0	0	0	3	1	3	1
max	3	10	2	2	10	0	0	0	3	3	3	3

Table 7.15: Constant Parameter Levels – Only Renewable Resources

Once again, we observe an increase of the computation times with the number of jobs and the resource factor. Moreover, the average computation times increase with a decreasing resource strength.

7.2 Truncated Exact Methods

In this section we briefly report the results we obtained by using the algorithm as a truncated exact method. That is, employing the variants P1,...,P8 (cf. Section 7.1) and the variant V6 with priority P1 we allowed ten seconds of CPU-time to solve the problem. We have made use of the instances generated for the evaluation and validation of ProGen (cf. Section 7.1). Recall, only 536 of the 640 instances have a feasible solution. Table 7.18 lists the results, where optimal refers to the number of problems which have been solved to optimality, best (worst) indicates the number of times where the calculated (feasible) solution has the best (worst) objective function

RF_R	RS_R	J	Feas.	V7	V8	Factor	J	Feas.	V7	V8	Factor
0.50	0.25	10	5	0.90	0.43	2.09	12	7	1.12	0.87	1.29
0.50	0.50		10	0.17	0.13	1.31		10	0.20	0.13	1.54
0.50	0.75		10	0.05	0.04	1.25		10	0.09	0.06	1.50
0.50	1.00		10	0.03	0.03	1.00		10	0.09	0.06	1.50
1.00	0.25	10	10	11.33	2.28	4.97	12	10	64.32	17.93	3.59
1.00	0.50		10	0.51	0.30	1.70		10	1.32	0.77	1.71
1.00	0.75		10	0.04	0.05	0.8		10	0.97	0.60	1.62
1.00	1.00		10	0.03	0.03	1.0		10	0.09	0.05	1.80
			75	1.68	0.41	4.10		77	8.81	2.62	3.36

Table 7.16: Computational Results - Bounding Rule 5 – J = 10, 12 (80486)

RF_R	RS_R	J	Feas.	V7	V8	Factor	J	Feas.	V7	V8	Factor
0.50	0.25	14	9	14.23	6.98	2.04	16	9	33.17	15.67	2.12
0.50	0.50		10	0.44	0.32	1.36		10	4.06	1.72	2.36
0.50	0.75		10	0.31	0.20	1.55		10	0.57	0.38	1.50
0.50	1.00		10	0.29	0.19	1.52		10	0.28	0.16	1.75
1.00	0.25	14	10	188.11	38.74	4.86	16	10	8456.79	1245.57	6.79
1.00	0.50		10	5.24	2.18	2.40		10	86.18	44.26	1.95
1.00	0.75		10	0.52	0.31	1.68		10	1.34	0.50	2.68
1.00	1.00		10	0.30	0.10	3.00		10	0.95	0.31	3.06
			79	26.33	6.12	4.30		79	1086.07	165.44	6.56

Table 7.17: Computational Results - Bounding Rule 5 – J = 14, 16 (80486)

value. The deviation is the average percentage deviation of the determined solution from the optimal solution. Solved indicates the number of problems, a feasible solution of which has been found.

	Variant								
	P1	P2	P3	P4	P5	P6	P7	P8	V6
Optimal	362	357	360	351	340	342	349	365	524
Best	363	359	362	353	342	344	352	366	529
Worst	51	60	61	63	72	63	55	46	4
Deviation[%]	4.691	5.012	5.316	5.570	5.089	5.304	4.668	4.539	0.226
Solved	474	472	481	476	467	471	470	474	535
Not solved	62	64	55	60	69	65	66	62	1

Table 7.18: Truncated Exact Methods - 10 sec. (80486)

There are only marginal differences between the priority rules. Clearly, the variant V6 outperforms all the others. 524 problems of the 536 instances have been solved to optimality. For only one problem an existing feasible solution could not be found. The average percentage deviation from optimality has been 0.226%.

Chapter 8

An Artificial Intelligence Approach

In this chapter we will use a logic programming approach in order to attack the scheduling problems described. Preliminary studies can be found in [46]. The outline is as follows:

In Section 8.1 we present reformulations of the model outlined in Chapter 1. Such reformulations make sense if e.g. the number of variables can be substantially reduced or if the computational tractability increases. A brief introduction to CHARME (cf. [16]), a PROLOG-based language, and the elements of the language can be found in Section 8.2. Computational results including a comparison with the general problem solver LINDO are provided in Section 8.3.

8.1 Model Reformulations

For each formulation of the model we consider two variants. First, we are dealing with the constraint satisfaction problem, that is, for the constraints imposed one or all feasible solutions have to be found. Second, we employ the makespan criterion as objective function, which has to be minimized.

We assume that the overall project has an unique source ($j = 1$) and an unique sink ($j = J$).

If the constraint satisfaction problem is considered, then for each project p, $p =$

$1, \ldots, P$, a project specific release date ρ_p and deadline δ_p has to be taken into account. We therefore assume the activities of project p to be consecutively labeled, that is, project p consists of the activities j, $j = FJ_p, \ldots, LJ_p$.

If the optimization problem is the point of attack then the upper bound $\overline{T}$ on the makespan is presumed to be given by the sum of the maximal durations of the jobs (cf. (1.1)). The release dates and deadlines are then skipped by defining $\rho_p := 0$ and $\delta_p := \overline{T}$, $p = 1, \ldots, P$.

$$\text{Minimize } \Phi(x) = \sum_{m=1}^{M_J} \sum_{t=EF_J}^{LF_J} t \cdot x_{Jmt} \tag{8.1}$$

s.t.

$$\sum_{m=1}^{M_j} \sum_{t=EF_j}^{LF_j} x_{jmt} = 1 \qquad j = 1, \ldots, J \tag{8.2}$$

$$\sum_{m=1}^{M_h} \sum_{t=EF_h}^{LF_h} t \cdot x_{hmt} \leq \sum_{m=1}^{M_j} \sum_{t=EF_j}^{LF_j} (t - d_{jm}) x_{jmt} \qquad j = 2, \ldots, J, h \in \mathcal{P}_j \tag{8.3}$$

$$\sum_{j=1}^{J} \sum_{m=1}^{M_j} k_{jmr}^{\rho} \sum_{q=t}^{t+d_{jm}-1} x_{jmq} \leq K_{rt}^{\rho} \qquad r \in R, t = 1, \ldots, \overline{T} \tag{8.4}$$

$$\sum_{j=1}^{J} \sum_{m=1}^{M_j} k_{jmr}^{\nu} \sum_{t=EF_j}^{LF_j} x_{jmt} \leq K_r^{\nu} \qquad r \in N \tag{8.5}$$

$$\sum_{m=1}^{M_j} \sum_{t=EF_j}^{LF_j} (t - d_{jm}) x_{jmt} \geq \rho_p \qquad p = 1, \ldots, P, j = FJ_p, \ldots, LJ_p \tag{8.6}$$

$$\sum_{m=1}^{M_j} \sum_{t=EF_j}^{LF_j} t \cdot x_{jmt} \leq \delta_p \qquad p = 1, \ldots, P, j = FJ_p, \ldots, LJ_p \tag{8.7}$$

$$x_{jmt} \in \{0, 1\} \qquad j = 1, \ldots, J, m = 1, \ldots, M_j, \quad t = 1, \ldots, \overline{T} \tag{8.8}$$

Table 8.1: Binary Programming Formulation (BINO)

With (BINO) we denote the optimization problem displayed in Table 8.1, where we define $\rho_p := 0$ and $\delta_p := \overline{T}$, $p = 1, \ldots, P$, and with (BINF) we refer to the related constraint satisfaction problem. We have augmented the model (cf. Table 1.3) by the constraints (8.6) and (8.7) in order to reflect the project specific release dates and deadlines.

We will now reformulate the problem (cf. [46]) by the use of binary variables x_{jm}, $j = 1, \ldots, J$, $m = 1, \ldots, M_j$,

$$x_{jm} = \begin{cases} 1 & , \text{ if job } j \text{ is performed in mode } m \\ 0 & , \text{ otherwise} \end{cases}$$

and integer variables CT_j, indicating the completion time of activity j, $j = 1, \ldots, J$. The model is displayed in Table 8.2. A_t denotes the set of jobs being active in period t, $t = 1, \ldots, \overline{T}$, i.e. (cf. [116])

$$A_t = \{j \in \{1, \ldots, J\}; CT_j + 1 - \sum_{m=1}^{M_j} d_{jm} x_{jm} \leq t \leq CT_j\}. \tag{8.9}$$

Again the objective function (8.10) is the makespan, which has to be minimized. The constraints (8.11) ensure that each job is assigned exactly one mode. The precedence relations are maintained by (8.12). By (8.13) we realize that the limited per period availability of the renewable resources is not exceeded. (8.14) ensures that nonrenewable resources are only consumed up to their availability. (8.15) and (8.16) ensure that the release dates and deadlines are met, respectively. We refer to the optimization problem with (MIPO), where the release dates and deadlines have to be skipped as described above. The constrained satisfaction problem is referred to as (MIPF).

Note, in comparison to the binary program formulation (BINO) the number of variables is substantially reduced. The number of variables in the binary programming formulation is a function of J, M and $\overline{T}$ ($O(J, M, \overline{T})$). The mixed integer programming formulation requires only $J + \sum_{j=1}^{J} M_j$ variables, which is $O(J, M)$.

We will now state an integer programming formulation employing integer variables to realize the mode selection. The model is outlined in Table 8.3. The decision

$$\text{Minimize } \Phi(x, CT) = CT_J \tag{8.10}$$

s.t.

$$\sum_{m=1}^{M_j} x_{jm} = 1 \qquad j = 1, \ldots, J \tag{8.11}$$

$$CT_h \leq CT_j - \sum_{m=1}^{M_j} d_{jm} x_{jm} \qquad j = 2, \ldots, J, h \in \mathcal{P}_j \tag{8.12}$$

$$\sum_{\substack{j=1 \\ j \in A_t}}^{J} \sum_{m=1}^{M_j} k_{jmr}^{\rho} x_{jm} \leq K_{rt}^{\rho} \qquad r \in R, t = 1, \ldots, \overline{T} \tag{8.13}$$

$$\sum_{j=1}^{J} \sum_{m=1}^{M_j} k_{jmr}^{\nu} x_{jm} \leq K_r^{\nu} \qquad r \in N \tag{8.14}$$

$$CT_j - \sum_{m=1}^{M_j} d_{jm} x_{jm} \geq \rho_p \qquad p = 1, \ldots, P, j = FJ_p, \ldots, LJ_p \tag{8.15}$$

$$CT_j \leq \delta_p \qquad p = 1, \ldots, P, j = FJ_p, \ldots, LJ_p \tag{8.16}$$

$$x_{jm} \in \{0, 1\} \qquad j = 1, \ldots, J, m = 1, \ldots, M_j \tag{8.17}$$

$$CT_j \in \{EF_j, \ldots, LF_j\} \qquad j = 1, \ldots, J \tag{8.18}$$

Table 8.2: Model Reformulation (MIPO)

variables CT_j represent the completion times of the activities j, $j = 1, \ldots, J$. Thus the objective function (8.19) is the makespan, which has to be minimized. The precedence constraints are maintained by (8.20). Limited per period and overall availability of the resources are taken into account by (8.21) and (8.22), respectively. The release dates and deadlines are enforced by (8.23) and (8.24). The mode selection is realized by the definition of the decision variables (8.25).

Only $2J$ integer variables are required in order to state the model. We refer to the optimization problem with (IPO) and the constraint satisfaction problem with (IPF). The adaptations for the optimization problem are again perfomed as described above.

Minimize $\Phi(CT, M) = CT_J$		(8.19)
s.t.		
$CT_h \leq CT_j - d_{jm_j}$	$j = 2, \ldots, J, h \in \mathcal{P}_j$	(8.20)
$\sum_{\substack{j=1 \\ CT_j - d_{jm_j}+1 \leq t \leq CT_j}}^{J} k^{\rho}_{jm_jr} \leq K^{\rho}_{rt}$	$r \in R, t = 1, \ldots, \overline{T}$	(8.21)
$\sum_{j=1}^{J} k^{\nu}_{jm_jr} \leq K^{\nu}_{r}$	$r \in \dot{N}$	(8.22)
$CT_j - d_{jm_j} \geq \rho_p$	$p = 1, \ldots, P, j = FJ_p, \ldots, LJ_p$	(8.23)
$CT_j \leq \delta_p$	$p = 1, \ldots, P, j = FJ_p, \ldots, LJ_p$	(8.24)
$m_j \in \{1, \ldots, M_j\}$	$j = 1, \ldots, J$	(8.25)
$CT_j \in \{EF_j, \ldots, LF_j\}$	$j = 1, \ldots, J$	(8.26)

Table 8.3: Model Reformulation (IPO)

8.2 A PROLOG-Based Implementation

In this section we will give a brief introduction to CHARME (cf. [16]), which is a PROLOG-based (cf. [21]) programming language. CHARME, a UNIX-based C-reimplemantation of CHIP (Constraint Handling in PROLOG, cf. [2]) has been especially designed for dealing with combinatorial (optimization) problems.

In contrast to imperative languages a CHARME program provides not a sequence of instructions describing how to solve a problem but a sequence of constraints describing what the problem is. It is a declarative language.

Given a certain number of variables, each of which can take a finite number of values (or equivalently, has a finite domain), the question is: Which values have to be assigned to the variables in order to meet a stated property. The property itself is described by a set of constraints which have to be met by a feasible solution.

CHARME's strength is it's ability of making use of the interactions between the constraints. Considering a constraint bearing any given variable, CHARME deduces information about the possible values the variable can take on. The new information is then compared with other constraints affected by the reduction of the domain of this variable. Loosely speaking, a constraint (hyper-) graph (cf. [10]) is constructed, where the nodes represent the variables and the arcs correspond to the constraints. The objective is to construct an arc- and path-consistent (cf. [78], [91]) constraint graph by making implicit constraints explicit.

In the artificial intelligence literature (cf. [29], [30], [76]) efficient search tree reduction techniques have attracted much attention. The main emphasis is on static (before variable instantiation) and dynamic (during variable instantiation) search tree reduction techniques.

In order to illustrate the formulation of the constraints we will reproduce some lines of code of the model (MIPO) making use of the high-level commands offered by CHARME. We define the input as follows (cf. [46]):

```
J             : number of jobs
MMax          : maximal number of modes per job
M[j]          : number of modes of job j, j=1,...,J
N             : number of nonrenewable resources
NRS[r]        : number of units available of nonrenewable
                resource r, r=1,...,N
NRR[j,m,r]    : number of units of resource r consumed by job j
                being performed in mode m, j=1,...,J,
                m=1,...,M[j], r=1,...,N
X[j,m]        : binary decision variable which is equal to one,
                if and only if job j is performed in mode m.
```

The constraints (8.11), that is, the assignment of exactly one mode to each job is realized by the following procedure:

```
define onemodeperjob(J,M,X)
{
for I in 1..J do
  sum(X[J,1..M[J]],1);
}
```

In order to reflect the constraints on the limited availability of the nonrenewable resources (8.14) we use the CHARME procedure "scal_prod" which stands for the inner product of two one-dimensional arrays.

```
define nonrenres(J,NRS,NRR,N,X)
{
 array Req :: [1..N]
 for I in 1..N do
   {
   Req[I] in 0..NRS[I];
   scal_prod(NRR[1..J,1..MMax,I]::[1..J*MMax],
   X[1..J,1..MMax]::[1..J*MMax],Req[I]);
   }
}
```

Since "scal_prod" is only applicable to one-dimensionl arrays, both two-dimensional arrays NRR[1..J,1..MMax,I][1] and X[1..J,1..MMax] have to be redimensioned.

8.3 Preliminary Computational Results

In this section we present the results of our computational study concerning the artificial intelligence approach. We compare the standard mixed integer programming problem solver LINDO with the realizations of the models outlined in Section 8.1 in CHARME.

[1]Note, fixing an index in a three-dimensional array and leaving the other indices free produces a two-dimensional array.

For our preliminary studies the examples used in the experiment have been manually constructed. We have considered single project problems. The constant parameter levels of the single project problems are displayed in Table 8.4. Ten problems have been generated.

	J	M_j	d_j	$\|R\|$	U_R	Q_R	$\|N\|$	U_N	Q_N	$\mathcal{S}_J$	$\mathcal{S}_j$	$\mathcal{P}_J$	$\mathcal{P}_j$
min	5	2	1	2	1	1	2	1	1	2	1	2	1
max	5	2	10	2	10	2	2	10	2	2	2	2	2

Table 8.4: Constant Parameter Levels

No.		1	2	3	4	5	6	7	8	9	10	μ_{CPU}
ρ_1		2	8	2	2	7	4	9	1	3	2	
δ_1		19	16	17	17	19	19	24	15	14	12	
Φ^*		17	6	17	16	11	16	13	12	10	12	
# feas. sol.		77	208	0	0	131	0	158	505	36	0	
$\overline{\Phi}$		16.16	6.00	14.42	13.00	11.00	13.38	13.00	12.00	9.18	8.40	
LINDO	R	42	9	26	23	31	11	22	62	39	23	28.8
	O	812	12	715	5004	334	756	370	1027	316	1057	1040.3
	1	74	60	18	602	8	80	33	9	68	66	101.8
BIN	O	–	–	–	–	–	–	–	–	–	–	–
	1	4	3	4	6	4	4	5	3	3	3	3.9
	A	14	28	4	6	22	4	27	72	7	3	18.7
MIP	O	246	14	4608	3233	44	20387	51	23	13	11338	3995.7
	1	0	0	52	29	0	173	0	0	0	25	27.9
	A	31	18	52	29	29	173	52	30	6	25	44.5
IP	O	574	1	10785	1034	797	122970	6173	243	1	440	14301.8
	1	0	0	3	8	0	72	6	0	0	9	9.8
	A	10	21	3	8	12	72	22	27	3	9	18.7

Table 8.5: Computational Results

Table 8.5 displays the solution times[2] and gives additional information about the

[2]Note, the LINDO-calculations have been performed on a PC-compatible computer with an

problems. The release date and deadline of the single project can be found in the second and third row, respectively. Surely, if only a single project is considered then the release date can be omitted by adapting the input data.

The optimal objective function value of the optimization problems and the number of feasible solutions of the constraint satisfaction problems are given in the fourth and fifth row, respectively. Row six shows the value of the LP-relaxations, that is the optimal objective function values of (BINO) if the integer conditions, the release date and deadline are omitted. These problems have been solved with LINDO. The related CPU-times are given in the seventh row (R). Moreover, we have solved the constraint satisfaction problem (BINF) and the optimization problem (BINO) with LINDO. In order to attack the problem (BINF) with LINDO we have used a constant (zero) function as objective function, that is, all the feasible solutions induce the same and optimal objective function value. The CPU-times for the determination of a makespan minimal schedule (O) and one feasible solution (1) by the use of LINDO is given in row eight and nine.

We have used CHARME to solve the constraint satisfaction problems (BINF), (MIPF) and (IPF), where for each formulation of the problem the times for the determination of one (1) and all (A) feasible solutions are displayed in the table. The optimization problems (BINO), (MIPO) and (IPO) have also been solved (O). Note, the differences between the computation times for the determination of all feasible solutions and one optimal solution are reasoned by the quality of the bound employed to find the latter one. We have used the sum of the maximum duration of the activities as upper bound. Due to limited availability of memory none of the instances has been solvable in the formulation (BINO). We can summarize Table 8.5 as follows:

80386dx processor, 40MHz clockpulse under DOS operating system. The CHARME experiment has been performed on an IBM PS2/95 system with an XP486 processor, 33MHz clockpulse under SCO-Unix operating system. Due to the comparison of the machines presented in Chapter 7 the latter is approximately three times faster than the former. A computation time of zero seconds indicates that less than half a second is required for solving the problem.

- With respect to average CPU-time the optimal solution has been obtained with LINDO more quickly than with the logic programming formulations. Although, there are problems which have been solved with lower effort by the CHARME formulations.

- If only one feasible solution has to be determined (1), then the binary programming fomulation (BINF) in CHARME outperforms all the other realizations with respect to average CPU-time. On the basis of the comparison factor of the machines (cf. Section 7.1) none of the artificial intelligence approaches has a higher average solution time than the one required by LINDO.

- The formulation (MIP) is with respect to average CPU-time for solving the constrained satisfaction problem the worst among the CHARME programs.

That is, if an optimization problem is the point of attack than LINDO is superior to the logic programming formulations. Otherwise, if the constraint satisfaction problem is under consideration, then the CHARME programs outperform LINDO.

Chapter 9

Applications

Beyond the applications mentioned in the preface concerning the management of projects, the models and algorithm proposed can be used in production planning. With production planning we refer to the preparation of decisions concerning the production. According to the addressed horizon three levels of planning can be distinguished (cf. [53], [103], pp. 668):

First, the long term planning process is concerned with the decision about the products to be produced and the design of the production plant. The addressed horizon is between two and ten years.

Second, the medium term planning process involves the decision about the product mix and the quantity of units to be produced, where the latter one is most commonly defined by a completely satisfied demand. Workloads are controlled on an aggregated level and adjusted if required. The addressed horizon is between six and twenty-four month.

Third, the short term planning process has its responsibility in the determination of lot sizes, the release dates and the assignment of resources to the jobs to be performed.

The specific details heavily depend on the characteristics of the production system under consideration (cf. [103], [125]):

- Origin of demand (make-to-order or make-to-stock-production),

- production rate (small batch or mass production),
- product structure (single-level or multi-level production),
- facility design (job-shop- or flow-shop-principle (cf. Chapter 2)).

We are concerned with make-to-order and/or small batch production within the medium and/or short term planning stage. Make-to-order companies make only little use of standard products, it is produced according to a customer specification. The following examples can be found (cf. [4], [20], [77], [93]):

- building ships, power stations, bridges, factories, highways etc.,
- machine construction,
- research and development projects,
- introduction of electronical data systems,
- mass meetings (sports, politics etc.).

In response to a customer's enquiry a quotation including the price and the delivery lead time is given. Competetive prices and attractive (realistic) delivery lead times play a key role in winning an order. In order to maintain a reputation which will gain further orders it is necessary that the promised date of delivery is met. Especially, due to the non-repetitive character of the production process a detailed planning process is required in order to meet the former objective.

Therefore, it is necessary to have a tool which is a link between the production planning (medium term) and the shop-floor control (short-term). A new tool is the so-called electronic leitstand[1] (cf. [1], [63]). In order to generate schedules and control the production a leitstand makes use of the following information (cf.[1]):

[1]The notion leitstand originates from the German language, it is to be translated as a control post or command center. The definition is given in the Lexikon der Verfahrenstechnik (cf. [100], p. 324).

- Production orders (e.g. quantaties and due dates),

- production facilities (e.g. fixture and tooling availabilities, machine status and capacity),

- process specifications (e.g. routings, processing times and subsitute machines).

To sum it all up, a leitstand is the electronical counterpart of the scheduling boards that can be found in the production control departments. A leitstand has the following five major components (cf.[1], [45] and [63]):

- A graphical component capable of providing a visual representation of the actvities and the precedence relation within a network. Furthermore order- and resource-specific Gantt-charts can be displayed.

- A schedule editor for manually manipulating schedules (e.g. adding and deleting activities (orders), changing quantities or changing sequences of operations on the machines).

- A data-base-management system for accessing the information relevant to scheduling, e.g. state of the machines, availability of resources, state of the activities (not started, in process, finished) and bill of material. Furthermore a communication component enables data-interchange with other computer subsystems.

- An evaluation component for measuring the goodness of a schedule under consideration[2].

- A generation component which determines on the basis of the given data feasible or even optimal schedules.

[2]Note, if more than one objective function is considered for the evaluation, then a multi criteria approach should be used (cf. [35]). However, in the current situation only a few systems exclude inefficient schedules (cf.[45]).

A survey of realizations of the leitstand concept can be found in [1] and [45]. In the latter reference and additionally in [42] a detailed discription of the PRISMA-system (production improvement of small batch machine tool assembly) is given.
The problem one encounters in the development of the generation component is that of solving a problem of type GRCPSP or a related one. The generation component of the PRISMA-leitstand makes use of this type of model, where additionally project specific release and due dates are incorporated (cf. Chapter 8). The makespan criterion is chosen for the following reasons as objective (cf. [42] and [44]):

- Short lead times reduce the complexity of the problem and avoid chaos on the shop-floor.
- Terms on payment define for nearly the complete amount (90%) the date of delivery as date of payment. Short lead times reduce the work-in-progress inventory.
- The liability of the data concerning the planning process are the higher the nearer they are to the planning phase.
- Having time reserve increases the probability of meeting the promised due date and avoids liquidated damages.
- The earlier the resources are used and the projects are completed the higher is the probability of having resources available for further orders.

Since not seldomly up to 2000 activities have to be scheduled on a per week basis (cf. [42]) the exact methods for solving the scheduling problems are of limited use. Moreover, due to the uncertainty of data involved it would be useless to determine an optimal solution. Therefore, heuristics (cf. e.g. [28], [41], [43], [67], [70], [71] [117]) are employed to solve the problem. However, the algorithms proposed offer a tool for evaluating heuristic approaches.

Chapter 10

Conclusions

We have considered the multi-mode resource-constrained project scheduling problem. The problem has been stated as a mathematical programming formulation, the versatility of which has been illustrated by examples. This should encourage the reader to try his hand on embedding other models in the formulation outlined.

Formal definitions of different types of schedules mentioned in the literature have been presented. The classification offers a criterion on distincting different enumeration procedures proposed in the literature. Moreover, the examples of the original work (cf. [111]) points out problems occuring with the adaption of verbal definitions to a more general framework.

A Branch and Bound algorithm proposed in the literature has been completely restructured. In spite of the more general representation of the algorithm the readability has been substantially increased. The new representation opened the possibility to derive the notion of an i-partial schedule, the basic framework of the algorithm. The formal definitions bring the essential information of the enumeration procedure into focus. Using the definitions new bounding rules have been described and verified. This acceleration schemes highly increase the performance of the algorithms, the limitations of which are illustrated by examples.

Nevertheless, there are some points of attack left: First, clearly bounding rule 2 can be extended to a set of job/mode combinations. That is, if interchanging the sequence of scheduling of the job/mode combinations produces the same start times for all the combinations and sequences, then only one of the sequences has to be evaluated.

Second, the computational study has turned out a lack of bounding rules for a high RF_R (resource factor of renewable resources) and a low RS_R (resource strength of renewable resources). Nevertheless, the relaxations of the renewable resource constraints (cf. Section 5.5 and 7.1) can probably close the gap.

The instance generator ProGen has been presented and used in order to evaluate the priority rules and acceleration schemes. The generator mainly makes use of

- the definition of a network in order to construct the precedence relations of the project,
- the resource factor as a normalized measure of density of the matrix,
- the resource strength as a normalized measure for the degree of availability of the resources (right hand side),
- tolerances in order to control the deviations from the requirements.

Clearly, all the ideas can be used in other contexts, as e.g. the generalized assignment problem (cf. [97]).

Using ProGen we have been able to perform an extensive computational study. Moreover, the computational study presented in Chapter 7 and [68] point out the necessity of an instance generator allowing to identify easy and hard instances. From this we can deduce information about problem specific algorithmic treatment. Furthermore, benchmark instances can give information about the algorithmic tractability of the problem under consideration.

Clearly, before creating an instance generator a problem has to be identified and formulated as a model. Subsequently solution procedures have to be developed and compared with standard problem solvers. This is the underlying idea of the artificial intelligence approach. Whereas, a linear programming approach not seldomly requires serious effort for its realization, a logic programming approach can substantially reduce the effort for preliminary investigations. The code is more compact and readable.

Beside the approaches outlined above other reformulations and relaxations can be used to solve the problem or to derive bounds on the objective function value. In the following we summarize two of them (cf. [110]).

Let $\mathbf{R}$ denote the set of real numbers. We consider a linear programming problem. Let $A \in \mathbf{R}^{m\times n}$, $c \in \mathbf{R}^n$, $b \in \mathbf{R}^m$ and

$$\begin{aligned} (B) \quad & \text{minimize } c^T x \\ & s.t. \\ & Ax \leq b \\ & x \in \{0,1\}^n \end{aligned}$$

One can easily verify that every feasible solution of (B) is an extreme point of the LP-relaxation $(\overline{B})$. That is, if c is integer-valued then the following strategy can be applied: Let $\overline{\Phi}$ be an integer-valued lower bound on the objective function value. Then we add the contraint

$$c^T x = \overline{\Phi}$$

to the contraints of (B). If we find among the feasible extreme points of the modified problem an integer-valued one, then we are finished. Otherwise, we increment $\overline{\Phi}$ by one and check the extreme points again.

Unfortunately, the algorithms proposed in the literature can only find all the vertices of small sized (6×15) problems within a reasonable (900 seconds) amount of CPU-time on an IBM mainframe 8083 (cf. [17] and [65]).

Moreover, a penalty approach (cf. [11]) can be chosen for solving a problem of the type GRCPSP. The underlying idea is the replacement of the problem at hand by a sequence of (easier) problems: We denote the objective funtion of the GRCPSP with $\Phi(x) = c^T x$. Using $N_j := \{(j,m,t); m = 1,\ldots,M_j, t = EF_j,\ldots,LF_j\}$, $j = 1,\ldots,J$, the constraints (1.7) can be depicted as

$$\sum_{i \in N_j} x_i = 1, \qquad j = 1, \ldots, J$$

with mutually disjoint sets $N_1, \ldots, N_J$. That is, we can choose the penalty function

$$g(x) = \sum_{j=1}^{J} (1 - \sum_{i \in N_j} x_i^2).$$

That is, with x fulfilling (1.7)-(1.10) and $n := \sum_{j=1}^{J} |N_j|$ it is

$$g(x) \begin{cases} = 0 & : \quad x \in \{0,1\}^n \\ > 0 & : \quad otherwise \end{cases}$$

One can verify that there is a $\overline{\lambda}$, $\overline{\lambda} > 0$, such that for every λ, $\lambda > \overline{\lambda}$ an optimal solution of

$$\begin{array}{lll} (B_\lambda) & \text{minimize } \Phi_\lambda(x) := c^T x + \lambda\, g(x) & \\ & s.t. & \\ & Ax \leq b & \\ & \sum_{i \in N_j} x_i = 1 & j = 1, \ldots, J \\ & x \in [0,1]^n & \end{array}$$

is an optimal solution of (B).

The resulting quadratic programming problem can be solved via convex envelope and Branch-and-Bound (cf. e.g. [62]). Since the approach requires serious effort for solving a problem of the type (B_λ), it is at present not suitable for combinatorial problems as e.g. project scheduling.

Bibliography

[1] ADELSBERGER, H.H. AND J.J. KANET (1991): The leitstand – A new tool for computer-integrated manufacturing. Production and Inventory Management Journal, Vol. 32, No. 1, pp. 43-48.

[2] AGGOUN, A. AND N. BALDICEANU (1991): Overview of the CHIP Compilersystem. Discussion Paper, Eighth International Conference on Logic Programming, Paris.

[3] ALVAREZ-VALDES, R. AND J.M. TAMARIT (1989): Heuristic algorithms for resource-constrained project scheduling: A review and an empirical analysis. In: Slowinski, R. and J. Weglarz (Eds.): Advances in project scheduling. Elsevier, Amsterdam, pp. 113-134.

[4] BADIRU, A.B. (1988): Project Management in Manufacturing and High Technology Operations. Wiley, New York et al.

[5] BAKER, K.R. (1974): Introduction to sequencing and scheduling, Wiley, New York et al.

[6] BALAS, E. (1971): Project scheduling with resource constraints. In: Beale, E.M.L. (Ed.): Applications of mathematical programming techniques. The English Universities Press, London, pp. 187-200.

[7] BARTUSCH, M.; R.H. MÖHRING AND F.J. RADERMACHER (1988): Scheduling project networks with resource constraints and time windows. Annals of Operations Research, Vol. 16, pp. 201-240.

[8] BELL, C.E. AND J. HAN (1991): A new heuristic solution method in resource-constrained project scheduling. Naval Research Logistics, Vol. 38, pp. 315-331.

[9] BELL, C.E. AND K. PARK (1990): Solving resource-constrained project scheduling problems by A* search. Naval Research Logistics, Vol. 37, pp. 61-84.

[10] BERGE, C. (1973): Graphs and Hypergraphs. North Holland, Amsterdam.

[11] BERTSEKAS, D. (1982): Constrained Optimization and Lagrange Multiplier Methods. Academic Press, New York.

[12] BLAZEWICZ, J.; K. ECKER; G. SCHMIDT AND J. WEGLARZ (1993): Scheduling in Computer and Manufacturing Systems. Springer, Berlin et al.

[13] BOCK, D.B. AND J.H. PATTERSON (1990): A comparison of due date setting, resource assignment, and job preemption heuristics for the multiproject scheduling problem. Decision Sciences, Vol. 21, pp. 387-402.

[14] BOCTOR, F.F. (1992): Heuristics for scheduling projects with resource restrictions and several resource-duration modes. Working Paper, Pavillon des Sciences de l'Administration, Universite Laval, Quebec, Canada.

[15] BOWMAN, E.H. (1959): The schedule-sequencing problem. Operations Research, Vol. 7, pp. 621-624.

[16] BULL (1990): Artificial Intelligence, Users Guide and Reference Manual. Bull S.A., Cedoc-Dilog.

[17] CHEN, P.C. AND P. HANSEN (1991): On-line and off-line vertex enumeration by adjacency lists. Operations Research Letters 10, pp. 403-409.

[18] CHRISTOFIDES, N.; R. ALVAREZ-VALDES AND J.M. TAMARIT (1987): Project scheduling with resource constraints: A branch and bound approach. European Journal of Operational Research, Vol. 29, pp. 262-273.

[19] CLARK, C.E. (1962): The PERT model for the distribution of an activity. Operations Research, Vol. 10, pp. 405-406.

[20] CLELAND, D.I. AND W.R. KING (1983): Systems Analysis and Project Management, 3rd edition. McGraw-Hill, New York et al.

[21] CLOCKSIN, W.F. AND C.S. MELLISH (1987): Programming in Prolog, 3rd edition. Springer, Berlin et al.

[22] CONWAY, R.W.; W.L. MAXWELL AND L.W. MILLER (1967): Theory of scheduling, Addison-Wesley, Reading, Massachusetts.

[23] COOPER, D.F. (1976): Heuristics for scheduling resource-constrained projects: An experimental investigation. Management Science, Vol. 22, pp. 1186-1194.

[24] DAVIES, E.M. (1973): An experimental investigation of resource allocation in multi activity projects. Operational Research Quarterly, Vol. 24, pp. 587-591.

[25] DAVIS, E.W. (1968): An exact algorithm for the multiple constrained-resource project scheduling problem. PhD Dissertation, Yale University, New Haven, USA.

[26] DAVIS, E.W. (1975): Project network summary measures constrained-resource scheduling. AIIE Transactions, Vol. 7, pp. 132-142.

[27] DAVIS, E.W. AND G.E. HEIDORN (1971): An algorithm for optimal project scheduling under multiple resource constraints. Management Science, Vol. 17, pp. B803-B816.

[28] DAVIS, E.W. AND J.H. PATTERSON (1975): A comparison of heuristic and optimum solutions in resource-constrained project scheduling. Management Science, Vol. 21, pp. 944-955.

[29] DECHTER, R. (1989/90): Enhancement schemes for constraint processing. Artificial Intelligence, Vol. 41, pp. 273-312.

[30] DECHTER, R. AND J. PEARL (1988): Network-based heuristics for constrained satisfaction problems. Artificial Intelligence, Vol. 34, pp. 1-38.

[31] DECKRO, R.F. AND J.E. HEBERT (1989): Resource-constrained project crashing. OMEGA, Vol. 17, pp. 69-79.

[32] DEMEULEMEESTER, E. (1992): Optimal algorithms for various classes of multiple resource-constrained project scheduling problems. PhD Dissertation, Katholieke Universiteit Leuven, Belgium.

[33] DEMEULEMEESTER, E. AND W. HERROELEN (1992): A branch-and-bound procedure for the multiple resource-constrained project scheduling problem. Management Science, Vol. 38, pp. 1803-1818.

[34] DEMEULEMEESTER, E. AND W. HERROELEN (1992): An efficient optimal solution procedure for the preemptive resource-constrained project scheduling problem, Working Paper, Department of Applied Economic Sciences, Katholieke Universiteit Leuven, Belgium.

[35] DINKELBACH, W. (1982): Entscheidungsmodelle. De Gruyter, New York.

[36] DINCBAS, M.; P. VAN HENTENRYCK; H. SIMONIS; A. AGGOUN; T. GRAF AND F. BERTHIER (1988): The constraint logic programming language CHIP. In: ICOT (Ed.): Proceedings on the Fifth Generation Computer Systems, Tokyo, pp. 693-702.

[37] DOMSCHKE, W. AND A. DREXL (1991): Kapazitätsplanung in Netzwerken — Ein Überblick über neuere Modelle und Verfahren. OR Spektrum, Bd. 13, pp. 63-76.

[38] DOMSCHKE, W. AND A. DREXL (1992): Einführung in Operations Research, 2nd edition. Springer, Berlin et al.

[39] DOMSCHKE, W.; A. SCHOLL AND S. VOSS (1993): Produktionsplanung – Ablauforganisatorische Aspekte. Springer, Berlin et al.

[40] DREXL, A. (1990): Fließbandaustaktung, Maschinenbelegung und Kapazitätsplanung in Netzwerken – Ein integrierender Ansatz. Zeitschrift für Betriebswirtschaft, Jg. 60, pp. 53-70.

[41] DREXL, A. (1991): Scheduling of project networks by job assignment. Management Science, Vol. 37, pp. 1590-1602.

[42] DREXL, A.; W. EVERSHEIM; R. GREMPE AND H. ESSER (1992): CIM im Werkzeugmaschinenbau: Der PRISMA-Montageleitstand. Manuskripte aus den Instituten für Betriebswirtschaftslehre, No. 295, Kiel.

[43] DREXL, A. AND J. GRÜNEWALD (1992): Nonpreemptive multi-mode resource-constrained project scheduling. IIE Transactions, to appear.

[44] DREXL, A. AND R. KOLISCH (1993): Produktionsplanung und -steuerung bei Einzel- und Kleinserienfertigung. Wirtschaftswissenschaftliches Studium, 22. Jg., Heft 2, pp. 60-66 and pp. 102-103.

[45] DREXL, A. AND R. KOLISCH (1993): Systementwicklung. Produktionsplanung und -steuerung bei Einzel- und Kleinserienfertigung: Leistandskonzepte. Wirtschaftswissenschaftliches Studium, 22. Jg., Heft 3, pp. 137-141.

[46] DREXL, A. AND A. SPRECHER (1993): Resource- and time window-constraint production scheduling with alternative process plans: An artifi-

cial intelligence approach. In: Fandel, G.; Th. Gulledge and A. Jones (Eds.): Operations Research in Production Planning and Control. Springer, Berlin et al., pp. 307-320.

[47] DRISCOL, J. AND A.A.A. ABDEL-SHAFI (1985): A simulation approach to evaluating assembly line balancing solutions. International Journal of Production Research, Vol. 23, pp. 975-985.

[48] ELMAGHRABY, S.E. (1967): On the expected duration of PERT type networks. Management Science, Vol. 13, pp. 299-306.

[49] ELMAGHRABY, S.E. (1977): Activity networks: Project planning and control by network models. Wiley, New York.

[50] ELMAGHRABY, S.E. (1989): The estimation of some network parameters in the PERT model of activity networks: Review and critique. In: Slowinski, R. and J. Weglarz (Eds.): Advances in project scheduling. Elsevier, Amsterdam, pp. 371-432.

[51] ELMAGHRABY, S.E. AND W.S. HERROELEN (1980): On the measurement of complexity in activity networks. European Journal of Operational Research, Vol. 5, pp. 223-234.

[52] FARNUM, N.R. AND L.W. STANTON (1987): Some results concerning the estimation of beta distribution parameters in PERT. Journal of the Operational Research Society, Vol. 38, pp. 287-290.

[53] FLEISCHMANN, G. (1988): Operations-Research-Modelle und -Verfahren in der Produktionsplanung. Zeitschrift für Betriebswirtschaft, Jg. 58, pp. 347-372.

[54] FRENCH, S. (1982): Sequencing and scheduling: An Introduction to the mathematics of the job-shop. Wiley, New York.

[55] GAREY, M.R. AND D.S. JOHNSON (1979): Computers and intractability: A Guide to the Theory of NP-Completeness. Freeman, San Francisco, CA.

[56] GOLENKO-GINZBURG, D. (1988): On the distribution of activity time in PERT. Journal of the Operational Research Society, Vol. 39, pp. 767-771.

[57] GONGUET, L. (1969): Comparison of three heuristic procedures for allocating resources and producing schedule. In: Lombaers, H.J.M. (Ed.): Project planning by network analysis, North-Holland, Amsterdam, pp. 249-255.

[58] HERROELEN, W.; E. DEMEULEMEESTER AND B.DODIN (1989): The generation of strongly-random activity networks. Working Paper, Department of Applied Economic Sciences, Katholieke Universiteit Leuven, Belgium.

[59] JACKSON, H.F.; P.T. BOGGS; S.G. NASH AND S. POWELL (1991): Guidelines for reporting results of computational experiments. Report of the ad hoc committee. Mathematical Programming, Vol. 49, pp. 413-425.

[60] JOHNSON, T.J.R. (1967): An algorithm for the resource-constrained project scheduling problem. PhD Dissertation, Massachusets Institute of Technology, USA.

[61] KAIMANN, R.A. (1974): Coefficients of network complexity. Management Science, Vol. 21, pp. 172-177.

[62] KALANTARI, B. AND J.B. ROSEN (1987): An algorithm for global minimization of linearly constrained concave quadratic functions. Mathematics of Operations Research, Vol. 12, No. 3, pp. 544-561.

[63] KANET, J.J. AND V. SRIDHARAN (1990): The electronic leitstand: A new tool for job scheduling. Manufactoring Review, Vol. 3, pp. 161-170.

[64] KARP, R.M. (1972): Reducibility among combinatorial problems. In: Miller, R.E. and J.W. Thatcher (Eds.): Complexity of computer applications. Plenum Press, New York, pp. 85-104.

[65] KHANG, D.B. AND O. FUJIWARA (1989): A new algorithm to find all vertices of a polytope. Operations Research Letters 8, pp. 261-264.

[66] KIM, S. AND R.C. LEACHMAN (1990): A hierarchical approach to multi-resource multi-project scheduling with explicit lateness costs. IIE Transactions, to appear.

[67] KOLISCH, R. (1993): Heuristic Algorithms for the Resource-Constrained Project Scheduling Problem. University Kiel, in preparation.

[68] KOLISCH, R.; A. SPRECHER AND A. DREXL (1992): Characterization and generation of a general class of resource-constrained project scheduling problems: Easy and hard instances. Manuskripte aus den Instituten für Betriebswirtschaftslehre, No. 301, Kiel.

[69] KURTULUS, I.S. (1983): Multi–project scheduling: Analysis of project performance. Working Paper, School of Business, VCU, Richmond, USA.

[70] KURTULUS, I.S. AND E.W. DAVIS (1982): Multi-project scheduling: Categorization of heuristic rules performance. Management Science, Vol. 28, pp. 161-172.

[71] KURTULUS, I.S. AND S.C. NARULA (1985): Multi-project scheduling: Analysis of project performance. IIE Transactions, Vol. 17, pp. 58-66.

[72] LAWRENCE, S.R. AND T.E. MORTON (1991): Resource-constrained multi-project scheduling with tardy costs: Comparing myopic, bottleneck, and resource pricing heuristics. Working Paper, Graduate School of Industrial Administration, Carnegie Mellon University, Pittsburgh, USA.

[73] MALCOLM, D.G.; J.H. ROSEBOOM; C.E. CLARK AND W. FAZAR (1959): Application of a technique for research and developement program evaluation. Operations Research, Vol. 7, pp. 646-669.

[74] MASON, A.T. AND C.L. MOODIE (1971): A branch and bound algorithm for minimizing cost in project scheduling. Management Science, Vol. 18, pp. B158-B173.

[75] MASTOR, A.A. (1970): An experimental investigation and comparative evaluation of production line balancing techniques. Management Science, Vol. 16, pp. 728-746.

[76] MESEGUER, P. (1989): Constraint satisfaction problems: An overview. AI-COM, Vol. 2, pp. 3-17.

[77] MODER J.J. AND C.R. PHILLIPS (1970): Project Management with PERT and CPM, 2nd edition. Van Nostrand Reinhold Company, New York et al.

[78] MOHR, R. AND T.C. HENDERSON (1986): Arc and path consistency. Artificial Intelligence, Vol. 28, pp. 225-233.

[79] NEUMANN, K. (1975): Operations Research Verfahren, Bd. 3. Hanser, München-Wien.

[80] NEUMANN, K. (1989): Scheduling of stochastic projects by means of GERT networks. In: Slowinski, R. and J. Weglarz (Eds.): Advances in project scheduling. Elsevier, Amsterdam, pp. 467-496.

[81] NEUMANN, K. (1990): Stochastic project networks — temporal analysis, scheduling and cost minimization. Springer, Berlin et al.

[82] PASCOE, T.L. (1966): Allocation of resources C.P.M. Revue Francaise Recherche Operationelle, No. 38, pp. 31-38.

[83] PATTERSON, J.H. (1976): Project scheduling: The effects of problem structure on heuristic performance. Naval Research Logistics Quarterly, Vol. 23, pp. 95-123.

[84] PATTERSON, J.H. (1984): A comparison of exact approaches for solving the multiple constrained resource, project scheduling problem. Management Science, Vol. 30, pp. 854-867.

[85] PATTERSON, J.H. AND J.J. ALBRACHT (1975): Assembly-line balancing: Zero-one programming with Fibonacci search. Operations Research, Vol. 23, pp. 166-172.

[86] PATTERSON, J.H. AND W.D. HUBER (1974): A horizon-varying, zero-one approach to project scheduling. Management Science, Vol. 20, pp. 990-998.

[87] PATTERSON, J.H. AND G.W. ROTH (1976): Scheduling a project under multiple resource constraints: A zero-one programming approach. AIIE Transactions, Vol. 8, pp. 449-455.

[88] PATTERSON, J.H.; R. SLOWINSKI; F.B. TALBOT AND J. WEGLARZ (1989): An algorithm for a general class of precedence and resource constrained scheduling problems. In: Slowinski, R. and J. Weglarz (Eds.): Advances in project scheduling. Elsevier, Amsterdam, pp. 3-28.

[89] PATTERSON, J.H.; R. SLOWINSKI; F.B. TALBOT AND J. WEGLARZ (1990): Computational experience with a backtracking algorithm for solving a general class of precedence and resource-constrained scheduling problems. European Journal of Operational Research, Vol. 49, pp. 68-79.

[90] PERT (1958): Program Evaluation Research Task. Phase 1, Summary Report, Special Projects Office, Bureau of Ordonance, Department of the Navy, Washington.

[91] PESCH, E.; A. DREXL AND A. KOLEN (1993): Automatisierte Wissensakquisition mittels modellbasierter Inferenz in CHARME. Discussion Paper, Universities of Maastricht and Kiel.

[92] PRITSKER, A.A.B. AND W.W. HAPP (1966): GERT: Graphical evaluation and review technique. Part I. Fundamentals. Journal of Industrial Engeneering, Vol. 17, pp. 267-274.

[93] PRITSKER, A.A.B. AND C.E. SIGAL (1983): Management decision making: a network simulation approach. Prentice Hall, Englewood Cliffs (N.J.).

[94] PRITSKER, A.A.B.; W.D. WATTERS AND P.M. WOLFE (1969): Multiproject scheduling with limited resources: A zero-one programming approach. Management Science, Vol. 16, pp. 93-108.

[95] RADERMACHER, F.J. (1985/86): Scheduling of project networks. Annals of Operations Research, Vol. 4, pp. 227-252.

[96] RINNOOY KAN, A.H.G. (1976): Machine scheduling problems: Classification, complexity and computation. Nijhoff, The Hague.

[97] ROSS, G.T. AND R.M. SOLAND (1975): A Branch and Bound algorithm for the generalized assignment problem. Mathematical Programming, Vol. 8, pp. 91-103.

[98] RUSSELL, R.A. (1986): A comparison of heuristics for scheduling projects with cash flows and resource restrictions. Management Science, Vol. 32, pp. 1291-1300.

[99] SAMPSON, S.E. AND E.N. WEISS (1992): Local search techniques for the resource-constrained project scheduling problem. Research Report, The Darden School, University of Virginia, USA.

[100] SCHIEFER, K. (1970): Lexikon der Verfahrenstechnik, Vol. 16. Deutsche Verlagsanstalt, Stuttgart.

[101] SCHRAGE, L. (1971): Solving resource-constrained network problems by implicit enumeration - nonpreemptive case. Operations Research, Vol. 18, pp. 263-278.

[102] SCHRAGE, L. (1979): A more portable Fortran random number generator. ACM Transactions on Mathematical Software, Vol. 5, pp. 132-138.

[103] SILVER, E.A. AND R. PETERSON (1985): Decision systems for inventory management and production planning, 2nd edition. Wiley, New York.

[104] SLOWINSKI, R. (1978): A node ordering heuristic for network scheduling under multiple resource constraints. Foundations of Control Engineering, Vol. 3, pp. 19-27.

[105] SLOWINSKI, R. (1980): Two approaches to problems of resource allocation among project activities: A comparative study. Journal of the Operational Research Society, Vol. 31, pp. 711-723.

[106] SLOWINSKI, R. (1981): Multiobjective network scheduling with efficient use of renewable and nonrenewable resources. European Journal of Operational Research, Vol. 7, pp. 265-273.

[107] SLOWINSKI, R. (1989): Multiobjective project scheduling under multiple-category resource constraints. In: Slowinski, R. and J. Weglarz (Eds.): Advances in project scheduling. Elsevier, Amsterdam, pp. 151-167.

[108] SMITH-DANIELS, D.E. AND V.L. SMITH-DANIELS (1987): Optimal project scheduling with materials ordering. IIE Transactions, Vol. 19, pp. 122-129.

[109] SPERANZA, M.G. AND C. VERCELLIS (1993): Hirarchical models for multi-project planning and scheduling. European Journal of Operational Research, Vol. 64, pp. 312-325.

[110] SPRECHER, A. AND A. DREXL (1992): Relaxations of the multi-mode resource-constrained project scheduling problem. Discussion Paper, presented at EURO/TIMS conference, Helsinki.

[111] SPRECHER, A.; R. KOLISCH AND A. DREXL (1993): Semi-active, active and non-delay schedules for the resource-constrained project scheduling problem. Manuskripte aus den Instituten für Betriebswirtschaftslehre, No. 307, Kiel.

[112] STINSON, J.P. (1976): A Branch and Bound algorithm for a general class of multiple resource-constrained scheduling problems. PhD Dissertation, Graduate School of Business Administration, University of North Carolina, USA.

[113] STINSON, J.P.; E.W. DAVIS AND B.M. KHUMAWALA (1978): Multiple resource-constrained scheduling using branch and bound. AIIE Transactions, Vol. 10, pp. 252-259.

[114] TALBOT, F.B. (1980): Project scheduling with resource-duration interactions: The nonpreemptive case. Working Paper, The Graduate School of Business Administration, University of Michigan, USA.

[115] TALBOT, F.B. (1982): Resource-constrained project scheduling with time-resource tradeoffs: The nonpreemptive case. Management Science, Vol. 28, pp. 1197-1210.

[116] TALBOT, F.B. AND J.H. PATTERSON (1978): An efficient integer programming algorithm with network cuts for solving resource-constrained scheduling problems. Management Science, Vol. 24, pp. 1163-1174.

[117] THESEN, A. (1976): Heuristic scheduling of activities under resource and precedence restrictions. Management Science, Vol. 23, pp. 412-422.

[118] THESEN, A. (1977): Measures of the restrictiveness of project networks. Networks, Vol. 7, pp. 193-208.

[119] ULUSOY, G. AND L. ÖZDAMAR (1989): Heuristic performance and network/resource characteristics in resource-constrained project scheduling. Journal of the Operational Research Society, Vol. 40, pp. 1145-1152.

[120] WEGLARZ, J. (1979): Project scheduling with discrete and continuous resources. IEEE Transactions on Systems, Man, and Cybernetics, Vol. 9, pp. 644-650.

[121] WEGLARZ, J. (1980): On certain models of resource allocation problems. Kybernetics, Vol. 9, pp. 61-66.

[122] WHITEHOUSE, G.E. (1973): Systems analysis and design using network techniques. Prentice Hall, Englewood Cliffs (N.J.).

[123] WIEST, J.D. (1964): Some properties of schedules for large projects with limited resources. Operations Research, Vol. 12, pp. 395-418.

[124] YAU, C. AND E. RITCHIE (1988): A linear model for estimating project resource levels and target completion times. Journal of the Operational Research Society, Vol. 39, pp. 855-866.

[125] ZÄPFEL, G. (1982): Produktionswirtschaft – Operatives Produktions-Management. De Gruyter, Berlin et al.

List of Figures

List of Tables

Springer-Verlag and the Environment

Lecture Notes in Economics and Mathematical Systems

For information about Vols. 1–234
please contact your bookseller or Springer-Verlag

Vol. 235: Stochastic Models in Reliability Theory Proceedings, 1984. Edited by S. Osaki and Y. Hatoyama. VII, 212 pages. 1984.

Vol. 236: G. Gandolfo, P.C. Padoan, A Disequilibrium Model of Real and Financial Accumulation in an Open Economy. VI, 172 pages. 1984.

Vol. 237: Misspecification Analysis. Proceedings, 1983. Edited by T.K. Dijkstra. V, 129 pages. 1984.

Vol. 238: W. Domschke, A. Drexl, Location and Layout Planning. IV, 134 pages. 1985.

Vol. 239: Microeconomic Models of Housing Markets. Edited by K. Stahl. VII, 197 pages. 1985.

Vol. 240: Contributions to Operations Research. Proceedings, 1984. Edited by K. Neumann and D. Pallaschke. V, 190 pages. 1985.

Vol. 241: U. Wittmann, Das Konzept rationaler Preiserwartungen. XI, 310 Seiten. 1985.

Vol. 242: Decision Making with Multiple Objectives. Proceedings, 1984. Edited by Y.Y. Haimes and V. Chankong. XI, 571 pages. 1985.

Vol. 243: Integer Programming and Related Areas. A Classified Bibliography 1981–1984. Edited by R. von Randow. XX, 386 pages. 1985.

Vol. 244: Advances in Equilibrium Theory. Proceedings, 1984. Edited by C.D. Aliprantis, O. Burkinshaw and N.J. Rothman. II, 235 pages. 1985.

Vol. 245: J.E.M. Wilhelm, Arbitrage Theory. VII, 114 pages. 1985.

Vol. 246: P.W. Otter, Dynamic Feature Space Modelling, Filtering and Self-Tuning Control of Stochastic Systems. XIV, 177 pages.1985.

Vol. 247: Optimization and Discrete Choice in Urban Systems. Proceedings, 1983. Edited by B.G. Hutchinson, P. Nijkamp and M. Batty Vl, 371 pages. 1985.

Vol. 248: Pural Rationality and Interactive Decision Processes. Proceedings, 1984. Edited by M. Grauer, M. Thompson and A.P. Wierzbicki. VI, 354 pages. 1985.

Vol. 249: Spatial Price Equilibrium: Advances in Theory, Computation and Application. Proceedings, 1984. Edited by P.T. Harker. VII, 277 pages. 1985.

Vol. 250: M. Roubens, Ph. Vincke, Preference Modelling. VIII, 94 pages. 1985.

Vol. 251: Input-Output Modeling. Proceedings, 1984. Edited by A. Smyshlyaev. VI, 261 pages. 1985.

Vol. 252: A. Birolini, On the Use of Stochastic Processes in Modeling Reliability Problems. VI, 105 pages. 1985.

Vol. 253: C. Withagen, Economic Theory and International Trade in Natural Exhaustible Resources. VI, 172 pages. 1985.

Vol. 254: S. Müller, Arbitrage Pricing of Contingent Claims. VIII, 151 pages. 1985.

Vol. 255: Nondifferentiable Optimization: Motivations and Applications. Proceedings, 1984. Edited by V.F. Demyanov and D. Pallaschke. VI, 350 pages. 1985.

Vol. 256: Convexity and Duality in Optimization. Proceedings, 1984. Edited by J. Ponstein. V, 142 pages. 1985.

Vol. 257: Dynamics of Macrosystems. Proceedings, 1984. Edited by J.-P. Aubin, D. Saari and K. Sigmund. VI, 280 pages. 1985.

Vol. 258: H. Funke, Eine allgemeine Theorie der Polypol- und Oligopolpreisbildung. III, 237 pages. 1985.

Vol. 259: Infinite Programming. Proceedings, 1984. Edited by E.J. Anderson and A.B. Philpott. XIV, 244 pages. 1985.

Vol. 260: H.-J. Kruse, Degeneracy Graphs and the Neighbourhood Problem. VIII, 128 pages. 1986.

Vol. 261: Th.R. Gulledge, Jr., N.K. Womer, The Economics of Made-to-Order Production. VI, 134 pages. 1986.

Vol. 262: H.U. Buhl, A Neo-Classical Theory of Distribution and Wealth. V, 146 pages. 1986.

Vol. 263: M. Schäfer, Resource Extraction and Market Struucture. XI, 154 pages. 1986.

Vol. 264: Models of Economic Dynamics. Proceedings, 1983. Edited by H.F. Sonnenschein. VII, 212 pages. 1986.

Vol. 265: Dynamic Games and Applications in Economics. Edited by T. Basar. IX, 288 pages. 1986.

Vol. 266: Multi-Stage Production Planning and Inventory Control. Edited by S. Axsäter, Ch. Schneeweiss and E. Silver. V, 264 pages.1986.

Vol. 267: R. Bemelmans, The Capacity Aspect of Inventories. IX, 165 pages. 1986.

Vol. 268: V. Firchau, Information Evaluation in Capital Markets. VII, 103 pages. 1986.

Vol. 269: A. Borglin, H. Keiding, Optimality in Infinite Horizon Economies. VI, 180 pages. 1986.

Vol. 270: Technological Change, Employment and Spatial Dynamics. Proceedings, 1985. Edited by P. Nijkamp. VII, 466 pages. 1986.

Vol. 271: C. Hildreth, The Cowles Commission in Chicago, 1939–1955. V, 176 pages. 1986.

Vol. 272: G. Clemenz, Credit Markets with Asymmetric Information. VIII,212 pages. 1986.

Vol. 273: Large-Scale Modelling and Interactive Decision Analysis. Proceedings, 1985. Edited by G. Fandel, M. Grauer, A. Kurzhanski and A.P. Wierzbicki. VII, 363 pages. 1986.

Vol. 274: W.K. Klein Haneveld, Duality in Stochastic Linear and Dynamic Programming. VII, 295 pages. 1986.

Vol. 275: Competition, Instability, and Nonlinear Cycles. Proceedings, 1985. Edited by W. Semmler. XII, 340 pages. 1986.

Vol. 276: M.R. Baye, D.A. Black, Consumer Behavior, Cost of Living Measures, and the Income Tax. VII, 119 pages. 1986.

Vol. 277: Studies in Austrian Capital Theory, Investment and Time. Edited by M. Faber. VI, 317 pages. 1986.

Vol. 278: W.E. Diewert, The Measurement of the Economic Benefits of Infrastructure Services. V, 202 pages. 1986.

Vol. 279: H.-J. Büttler, G. Frei and B. Schips, Estimation of Disequilibrium Modes. VI, 114 pages. 1986.

Vol. 280: H.T. Lau, Combinatorial Heuristic Algorithms with FORTRAN. VII, 126 pages. 1986.

Vol. 281: Ch.-L. Hwang, M.-J. Lin, Group Decision Making under Multiple Criteria. XI, 400 pages. 1987.

Vol. 282: K. Schittkowski, More Test Examples for Nonlinear Programming Codes. V, 261 pages. 1987.

Vol. 283: G. Gabisch, H.-W. Lorenz, Business Cycle Theory. VII, 229 pages. 1987.

Vol. 284: H. Lütkepohl, Forecasting Aggregated Vector ARMA Processes. X, 323 pages. 1987.

Vol. 285: Toward Interactive and Intelligent Decision Support Systems. Volume 1. Proceedings, 1986. Edited by Y. Sawaragi, K. Inoue and H. Nakayama. XII, 445 pages. 1987.

Vol. 286: Toward Interactive and Intelligent Decision Support Systems. Volume 2. Proceedings, 1986. Edited by Y. Sawaragi, K. Inoue and H. Nakayama. XII, 450 pages. 1987.

Vol. 287: Dynamical Systems. Proceedings, 1985. Edited by A.B. Kurzhanski and K. Sigmund. VI, 215 pages. 1987.

Vol. 288: G.D. Rudebusch, The Estimation of Macroeconomic Disequilibrium Models with Regime Classification Information. VII,128 pages. 1987.

Vol. 289: B.R. Meijboom, Planning in Decentralized Firms. X, 168 pages. 1987.

Vol. 290: D.A. Carlson, A. Haurie, Infinite Horizon Optimal Control. XI, 254 pages. 1987.

Vol. 291: N. Takahashi, Design of Adaptive Organizations. VI, 140 pages. 1987.

Vol. 292: I. Tchijov, L. Tomaszewicz (Eds.), Input-Output Modeling. Proceedings, 1985. VI, 195 pages. 1987.

Vol. 293: D. Batten, J. Casti, B. Johansson (Eds.), Economic Evolution and Structural Adjustment. Proceedings, 1985. VI, 382 pages.

Vol. 294: J. Jahn, W. Knabs (Eds.), Recent Advances and Historical Development of Vector Optimization. VII, 405 pages. 1987.

Vol. 295. H. Meister, The Purification Problem for Constrained Games with Incomplete Information. X, 127 pages. 1987.

Vol. 296: A. Börsch-Supan, Econometric Analysis of Discrete Choice. VIII, 211 pages. 1987.

Vol. 297: V. Fedorov, H. Läuter (Eds.), Model-Oriented Data Analysis. Proceedings, 1987. VI, 239 pages. 1988.

Vol. 298: S.H. Chew, Q. Zheng, Integral Global Optimization. VII, 179 pages. 1988.

Vol. 299: K. Marti, Descent Directions and Efficient Solutions in Discretely Distributed Stochastic Programs. XIV, 178 pages. 1988.

Vol. 300: U. Derigs, Programming in Networks and Graphs. XI, 315 pages. 1988.

Vol. 301: J. Kacprzyk, M. Roubens (Eds.), Non-Conventional Preference Relations in Decision Making. VII, 155 pages. 1988.

Vol. 302: H.A. Eiselt, G. Pederzoli (Eds.), Advances in Optimization and Control. Proceedings, 1986. VIII, 372 pages. 1988.

Vol. 303: F.X. Diebold, Empirical Modeling of Exchange Rate Dynamics. VII, 143 pages. 1988.

Vol. 304: A. Kurzhanski, K. Neumann, D. Pallaschke (Eds.), Optimization, Parallel Processing and Applications. Proceedings, 1987. VI, 292 pages. 1988.

Vol. 305: G.-J.C.Th. van Schijndel, Dynamic Firm and Investor Behaviour under Progressive Personal Taxation. X, 215 pages.1988.

Vol. 306: Ch. Klein, A Static Microeconomic Model of Pure Competition. VIII, 139 pages. 1988.

Vol. 307: T.K. Dijkstra (Ed.), On Model Uncertainty and its Statistical Implications. VII, 138 pages. 1988.

Vol. 308: J.R. Daduna, A. Wren (Eds.), Computer-Aided Transit Scheduling. VIII, 339 pages. 1988.

Vol. 309: G. Ricci, K. Velupillai (Eds.), Growth Cycles and Multisectoral Economics: the Goodwin Tradition. III, 126 pages. 1988.

Vol. 310: J. Kacprzyk, M. Fedrizzi (Eds.), Combining Fuzzy Imprecision with Probabilistic Uncertainty in Decision Making. IX, 399 pages. 1988.

Vol. 311: R. Färe, Fundamentals of Production Theory. IX, 163 pages. 1988.

Vol. 312: J. Krishnakumar, Estimation of Simultaneous Equation Models with Error Components Structure. X, 357 pages. 1988.

Vol. 313: W. Jammernegg, Sequential Binary Investment Decisions. VI, 156 pages. 1988.

Vol. 314: R. Tietz, W. Albers, R. Selten (Eds.), Bounded Rational Behavior in Experimental Games and Markets. VI, 368 pages. 1988.

Vol. 315: I. Orishimo, G.J.D. Hewings, P. Nijkamp (Eds), Information Technology: Social and Spatial Perspectives. Proceedings 1986. VI, 268 pages. 1988.

Vol. 316: R.L. Basmann, D.J. Slottje, K. Hayes, J.D. Johnson, D.J. Molina, The Generalized Fechner-Thurstone Direct Utility Function and Some of its Uses. VIII, 159 pages. 1988.

Vol. 317: L. Bianco, A. La Bella (Eds.), Freight Transport Planning and Logistics. Proceedings, 1987. X, 568 pages. 1988.

Vol. 318: T. Doup, Simplicial Algorithms on the Simplotope. VIII, 262 pages. 1988.

Vol. 319: D.T. Luc, Theory of Vector Optimization. VIII, 173 pages. 1989.

Vol. 320: D. van der Wijst, Financial Structure in Small Business. VII, 181 pages. 1989.

Vol. 321: M. Di Matteo, R.M. Goodwin, A. Vercelli (Eds.), Technological and Social Factors in Long Term Fluctuations. Proceedings. IX, 442 pages. 1989.

Vol. 322: T. Kollintzas (Ed.), The Rational Expectations Equilibrium Inventory Model. XI, 269 pages. 1989.

Vol. 323: M.B.M. de Koster, Capacity Oriented Analysis and Design of Production Systems. XII, 245 pages. 1989.

Vol. 324: I.M. Bomze, B.M. Pötscher, Game Theoretical Foundations of Evolutionary Stability. VI, 145 pages. 1989.

Vol. 325: P. Ferri, E. Greenberg, The Labor Market and Business Cycle Theories. X, 183 pages. 1989.

Vol. 326: Ch. Sauer, Alternative Theories of Output, Unemployment, and Inflation in Germany: 1960–1985. XIII, 206 pages. 1989.

Vol. 327: M. Tawada, Production Structure and International Trade. V, 132 pages. 1989.

Vol. 328: W. Güth, B. Kalkofen, Unique Solutions for Strategic Games. VII, 200 pages. 1989.

Vol. 329: G. Tillmann, Equity, Incentives, and Taxation. VI, 132 pages. 1989.

Vol. 330: P.M. Kort, Optimal Dynamic Investment Policies of a Value Maximizing Firm. VII, 185 pages. 1989.

Vol. 331: A. Lewandowski, A.P. Wierzbicki (Eds.), Aspiration Based Decision Support Systems. X, 400 pages. 1989.

Vol. 332: T.R. Gulledge, Jr., L.A. Litteral (Eds.), Cost Analysis Applications of Economics and Operations Research. Proceedings. VII, 422 pages. 1989.

Vol. 333: N. Dellaert, Production to Order. VII, 158 pages. 1989.

Vol. 334: H.-W. Lorenz, Nonlinear Dynamical Economics and Chaotic Motion. XI, 248 pages. 1989.

Vol. 335: A.G. Lockett, G. Islei (Eds.), Improving Decision Making in Organisations. Proceedings. IX, 606 pages. 1989.

Vol. 336: T. Puu, Nonlinear Economic Dynamics. VII, 119 pages. 1989.

Vol. 337: A. Lewandowski, I. Stanchev (Eds.), Methodology and Software for Interactive Decision Support. VIII, 309 pages. 1989.

Vol. 338: J.K. Ho, R.P. Sundarraj, DECOMP: an Implementation of Dantzig-Wolfe Decomposition for Linear Programming. VI, 206 pages.

Vol. 339: J. Terceiro Lomba, Estimation of Dynamic Econometric Models with Errors in Variables. VIII, 116 pages. 1990.

Vol. 340: T. Vasko, R. Ayres, L. Fontvieille (Eds.), Life Cycles and Long Waves. XIV, 293 pages. 1990.

Vol. 341: G.R. Uhlich, Descriptive Theories of Bargaining. IX, 165 pages. 1990.

Vol. 342: K. Okuguchi, F. Szidarovszky, The Theory of Oligopoly with Multi-Product Firms. V, 167 pages. 1990.

Vol. 343: C. Chiarella, The Elements of a Nonlinear Theory of Economic Dynamics. IX, 149 pages. 1990.

Vol. 344: K. Neumann, Stochastic Project Networks. XI, 237 pages. 1990.

Vol. 345: A. Cambini, E. Castagnoli, L. Martein, P Mazzoleni, S. Schaible (Eds.), Generalized Convexity and Fractional Programming with Economic Applications. Proceedings, 1988. VII, 361 pages. 1990.

Vol. 346: R. von Randow (Ed.), Integer Programming and Related Areas. A Classified Bibliography 1984–1987. XIII, 514 pages. 1990.

Vol. 347: D. Ríos Insua, Sensitivity Analysis in Multiobjective Decision Making. XI, 193 pages. 1990.

Vol. 348: H. Störmer, Binary Functions and their Applications. VIII, 151 pages. 1990.

Vol. 349: G.A. Pfann, Dynamic Modelling of Stochastic Demand for Manufacturing Employment. VI, 158 pages. 1990.

Vol. 350: W.-B. Zhang, Economic Dynamics. X, 232 pages. 1990.

Vol. 351: A. Lewandowski, V. Volkovich (Eds.), Multiobjective Problems of Mathematical Programming. Proceedings, 1988. VII, 315 pages. 1991.

Vol. 352: O. van Hilten, Optimal Firm Behaviour in the Context of Technological Progress and a Business Cycle. XII, 229 pages. 1991.

Vol. 353: G. Ricci (Ed.), Decision Processes in Economics. Proceedings, 1989. III, 209 pages 1991.

Vol. 354: M. Ivaldi, A Structural Analysis of Expectation Formation. XII, 230 pages. 1991.

Vol. 355: M. Salomon. Deterministic Lotsizing Models for Production Planning. VII, 158 pages. 1991.

Vol. 356: P. Korhonen, A. Lewandowski, J . Wallenius (Eds.), Multiple Criteria Decision Support. Proceedings, 1989. XII, 393 pages. 1991.

Vol. 357: P. Zörnig, Degeneracy Graphs and Simplex Cycling. XV, 194 pages. 1991.

Vol. 358: P. Knottnerus, Linear Models with Correlated Disturbances. VIII, 196 pages. 1991.

Vol. 359: E. de Jong, Exchange Rate Determination and Optimal Economic Policy Under Various Exchange Rate Regimes. VII, 270 pages. 1991.

Vol. 360: P. Stalder, Regime Translations, Spillovers and Buffer Stocks. VI, 193 pages . 1991.

Vol. 361: C. F. Daganzo, Logistics Systems Analysis. X, 321 pages. 1991.

Vol. 362: F. Gehrels, Essays In Macroeconomics of an Open Economy. VII, 183 pages. 1991.

Vol. 363: C. Puppe, Distorted Probabilities and Choice under Risk. VIII, 100 pages . 1991

Vol. 364: B. Horvath, Are Policy Variables Exogenous? XII, 162 pages. 1991.

Vol. 365: G. A. Heuer, U. Leopold-Wildburger. Balanced Silverman Games on General Discrete Sets. V, 140 pages. 1991.

Vol. 366: J. Gruber (Ed.), Econometric Decision Models. Proceedings, 1989. VIII, 636 pages. 1991.

Vol. 367: M. Grauer, D. B. Pressmar (Eds.), Parallel Computing and Mathematical Optimization. Proceedings. V, 208 pages. 1991.

Vol. 368: M. Fedrizzi, J. Kacprzyk, M. Roubens (Eds.), Interactive Fuzzy Optimization. VII, 216 pages. 1991.

Vol. 369: R. Koblo, The Visible Hand. VIII, 131 pages.1991.

Vol. 370: M. J. Beckmann, M. N. Gopalan, R. Subramanian (Eds.), Stochastic Processes and their Applications. Proceedings, 1990. XLI, 292 pages. 1991.

Vol. 371: A. Schmutzler, Flexibility and Adjustment to Information in Sequential Decision Problems. VIII, 198 pages. 1991.

Vol. 372: J. Esteban, The Social Viability of Money. X, 202 pages. 1991.

Vol. 373: A. Billot, Economic Theory of Fuzzy Equilibria. XIII, 164 pages. 1992.

Vol. 374: G. Pflug, U. Dieter (Eds.), Simulation and Optimization. Proceedings, 1990. X, 162 pages. 1992.

Vol. 375: S.-J. Chen, Ch.-L. Hwang, Fuzzy Multiple Attribute Decision Making. XII, 536 pages. 1992.

Vol. 376: K.-H. Jöckel, G. Rothe, W. Sendler (Eds.), Bootstrapping and Related Techniques. Proceedings, 1990. VIII, 247 pages. 1992.

Vol. 377: A. Villar, Operator Theorems with Applications to Distributive Problems and Equilibrium Models. XVI, 160 pages. 1992.

Vol. 378: W. Krabs, J. Zowe (Eds.), Modern Methods of Optimization. Proceedings, 1990. VIII, 348 pages. 1992.

Vol. 379: K. Marti (Ed.), Stochastic Optimization. Proceedings, 1990. VII, 182 pages. 1992.

Vol. 380: J. Odelstad, Invariance and Structural Dependence. XII, 245 pages. 1992.

Vol. 381: C. Giannini, Topics in Structural VAR Econometrics. XI, 131 pages. 1992.

Vol. 382: W. Oettli, D. Pallaschke (Eds.), Advances in Optimization. Proceedings, 1991. X, 527 pages. 1992.

Vol. 383: J. Vartiainen, Capital Accumulation in a Corporatist Economy. VII, 177 pages. 1992.

Vol. 384: A. Martina, Lectures on the Economic Theory of Taxation. XII, 313 pages. 1992.

Vol. 385: J. Gardeazabal, M. Regúlez, The Monetary Model of Exchange Rates and Cointegration. X, 194 pages. 1992.

Vol. 386: M. Desrochers, J.-M. Rousseau (Eds.), Computer-Aided Transit Scheduling. Proceedings, 1990. XIII, 432 pages. 1992.

Vol. 387: W. Gaertner, M. Klemisch-Ahlert, Social Choice and Bargaining Perspectives on Distributive Justice. VIII, 131 pages. 1992.

Vol. 388: D. Bartmann, M. J. Beckmann, Inventory Control. XV, 252 pages. 1992.

Vol. 389: B. Dutta, D. Mookherjee, T. Parthasarathy, T. Raghavan, D. Ray, S. Tijs (Eds.), Game Theory and Economic Applications. Proceedings, 1990. ??, ?? pages. 1992.

Vol. 390: G. Sorger, Minimum Impatience Theorem for Recursive Economic Models. X, 162 pages. 1992.

Vol. 391: C. Keser, Experimental Duopoly Markets with Demand Inertia. X, 150 pages. 1992.

Vol. 392: K. Frauendorfer, Stochastic Two-Stage Programming. VIII, 228 pages. 1992.

Vol. 393: B. Lucke, Price Stabilization on World Agricultural Markets. XI, 274 pages. 1992.

Vol. 394: Y.-J. Lai, C.-L. Hwang, Fuzzy Mathematical Programming. XIII, 301 pages. 1992.

Vol. 395: G. Haag, U. Mueller, K. G. Troitzsch (Eds.), Economic Evolution and Demographic Change. XVI, 409 pages. 1992.

Vol. 396: R. V. V. Vidal (Ed.), Applied Simulated Annealing. VIII, 358 pages. 1992.

Vol. 397: J. Wessels, A. P. Wierzbicki (Eds.), User-Oriented Methodology and Techniques of Decision Analysis and Support. Proceedings, 1991. XII, 295 pages. 1993.

Vol. 398: J.-P. Urbain, Exogeneity in Error Correction Models. XI, 189 pages. 1993.

Vol. 399: F. Gori, L. Geronazzo, M. Galeotti (Eds.), Nonlinear Dynamics in Economics and Social Sciences. Proceedings, 1991. VIII, 367 pages. 1993.

Vol. 400: H. Tanizaki, Nonlinear Filters. XII, 203 pages. 1993.

Vol. 401: K. Mosler, M. Scarsini, Stochastic Orders and Applications. V, 379 pages. 1993.

Vol. 402: A. van den Elzen, Adjustment Processes for Exchange Economies and Noncooperative Games. VII, 146 pages. 1993.

Vol. 403: G. Brennscheidt, Predictive Behavior. VI, 227 pages. 1993.

Vol. 404: Y.-J. Lai, Ch.-L. Hwang, Fuzzy Multiple Objective Decision Making. XIV, 475 pages. 1994.

Vol. 405: S. Komlósi, T. Rapcsák, S. Schaible (Eds.), Generalized Convexity. Proceedings, 1992. VIII, 404 pages. 1994.

Vol. 406: N. M. Hung, N. V. Quyen, Dynamic Timing Decisions Under Uncertainty. X, 194 pages. 1994.

Vol. 407: M. Ooms, Empirical Vector Autoregressive Modeling. XIII, 380 pages. 1994.

Vol. 408: K. Haase, Lotsizing and Scheduling for Production Planning. VIII, 118 pages. 1994.

Vol. 409: A. Sprecher, Resource-Constrained Project Scheduling. XII, 142 pages. 1994.

Vol. 410: R. Winkelmann, Count Data Models. XI, 213 pages. 1994.

Vol. 411: S. Dauzère-Péres, J.-B. Lasserre, An Integrated Approach in Production Planning and Scheduling. XVI, 137 pages. 1994.